530 Polkinghorne
P769p The particle play

Glendale College
Library

The Particle Play

High Energy Physics in a Swiss setting. In the foreground is the CERN Laboratory with the underground ring of the SPS indicated by the broken lines. Beyond is Geneva and the Mont Blanc massif (*photograph* – CERN).

The Particle Play

An account of the ultimate constituents of matter

J. C. POLKINGHORNE F.R.S.

University of Cambridge

W. H. Freeman and Company

Oxford and San Francisco

530
P769p

W. H. Freeman and Company Limited
20 Beaumont Street, Oxford, OX1 2NQ
660 Market Street, San Francisco, California, 94104

Library of Congress Cataloging in Publication Data

Polkinghorne, J C 1930–
The particle play.
Includes bibliographies and index.
1. Matter–Constitution. 2. Quarks.
I. Title.
QC173.P559 530 79–17846
ISBN 0–7167–1177–X

Copyright © 1979 by W. H. Freeman and Company Limited

No part of this book may be reproduced by any mechanical, photographic, or electronic process, or in the form of a phonographic recording, nor may it be stored in a retrieval system, transmitted, or otherwise copied for public or private use, without the written permission of the publisher.

Typeset by Universities Press, Belfast
Printed in Great Britain
by Fletcher and Son Ltd., Norwich

4/81

FOR RUTH

AND

PETER, ISOBEL

AND MICHAEL

Preface

It is natural to be curious about what things are made of. Enormous intellectual effort has been expended in the attempt to find an answer. Almost as substantial have been the sums of taxpayers' money spent in recent years on the elaborate apparatus necessary for the quest. I believe the double investment has been worthwhile and that the progress made represents an achievement of great cultural value. It has given enormous pleasure and excitement to those engaged on it, however minor their role. If this book can communicate a little of this exhilaration to the general reader and patient taxpayer, it will have served its purpose.

The task is not an easy one. The way things behave in the microworld is often very different from that which our experience of everyday life would lead us to expect. The methods for understanding the physics of the very small are centred on quantum mechanics and special relativity, theories which for their natural and perfect expression require more mathematics than the layman is likely to have at his command. I have chosen, in fact, to try to dispense with mathematics altogether (except for a moment of indulgence at the end of Chapter IV). Access to general ideas seemed more important than the exposition of their technical details.

Anyone who writes a book on science for a lay audience finds that he acquires an *alter ego*, who is a cousin of the internalized parent of psychoanalytic theory. I call mine 'the internalized colleague'. There are two messages that he whispers to me. One is 'You left that out'. Of course I did, for the choice of what to leave unsaid is certainly as important as the choice of what to put in, in a book of this kind. Some things get omitted because they are frankly too hard to convey in non-technical terms. Others because I am not writing an encyclopaedia and there are limits to the sheer quantity of material you can get across without your readers' eyes glazing over. For example, it is a matter of regret that the book does not give an account of the remarkable properties of the K^0–$\bar{K}^0$ system but I felt that to put it in would overload the relevant section.

The second message, whispered to me by my internalized

colleague is 'That isn't quite correct without adding this proviso and specifying that exception'. Here again I am unrepentant. I have tried to make the book as accurate as I can within the inescapably broad-brush terms of reference that must apply to a book which attempts a non-technical treatment. I cannot guarantee that every statement of mine would be recognized by a professional as true *au pied de la lettre*. However, I have made as strenuous efforts as I am able not to mislead.

I am very grateful to my wife for reading the manuscript, to Miss Wendy Knightley for skilfully typing it, to Mr. Chris Chalk for the drawings, and to CERN and DESY for the photographs. Above all I am grateful to Dr. Michael Rodgers. It was a suggestion from him that started me on the task of writing this book and he has made many other suggestions which substantially improved it. Of course the deficiencies which remain are my responsibility.

J. C. POLKINGHORNE

Department of Applied Mathematics and Theoretical Physics,
University of Cambridge.
March 1979

Contents

I. DRAMATIS PERSONAE
Page 1

II. STAGE MACHINERY
Page 25

III. SETTING THE SCENE
Page 36

IV. THE PLOT THICKENS
Page 48

V. DRAMATURGY
Page 69

VI. EXITS AND ENTRANCES
Page 82

VII. QUARKS REPRISE
Page 92

VIII. DENOUEMENT
Page 106

IX. FORTHCOMING ATTRACTIONS
Page 121

X. EPILOGUE
Page 124

GLOSSARY
Page 127

INDEX
Page 135

My soul, sit thou a patient looker-on;
Judge not the play before the play is done:
Her plot hath many changes: every day
Speaks a new scene; the last act crowns the play.

Francis Quarles (1592–1644)
Epigram. Respice Finem.

I

Dramatis Personae

The physical world is full of diverse objects. Since the time of the Greeks, people have wondered whether it could be the case that this bewildering variety was constructed out of a few simple elements. The pre-Socratic philosophers canvassed two alternative approaches to the problem. Some thought that there was a single universal substance and that the variety of its manifestations was due to the varying degrees of compactness with which it was present. Thus Anaximenes saw air as the primal material. Rarefy it and you had fire. Condense it and you get water and finally stone. Such a picture has a long intellectual pedigree. It was still current in Newton's time, when it was considered that the differing sizes of the particles into which universal matter was condensed accounted for the different properties of materials. However, the modern view has inclined to the second of the two approaches which were already being proposed over 2500 years ago.

In this second approach there are thought to be several different basic substances and the variety of observed matter derives from the varying proportions with which they are combined. The ancient system based on air, earth, fire, and water is an elegant example of such a theory. Of course we have had from time to time to revise our opinions about the nature of the basic constituents, but this point of view is the one which has dominated the study of the fundamental structure of matter over the last 150 years. The answer must be economic to be intellectually appealing. A plethora of fundamental objects is unsatisfying.

It is worth stopping a moment to consider the last statement. While such a plethora would indeed be unsatisfying, nature is nature and if that was the way things were then wouldn't we just have to make the best of it? Of course we would. However, it is interesting that such notions of economy and elegance, especially when expressed in mathematical form, have frequently proved valuable guides to a better understanding of the physical world. It is a recognized technique in elementary particle physics to seek theories which are compact and mathematically beautiful, in the expectation that they will then prove to be the ones realized in nature. This is a striking fact. Different people will make different interpretations of it. I shall explain at the end of this book how I view the question.

Progress in understanding the structure of matter is made in a spiral fashion. First some constituents are identified. Then further investigation reveals that a few more are required to do justice to the complexity of the phenomena. After a while, as their number increases, the situation becomes distinctly unaesthetic. A plethora is threatening. Just when all seems lost, a structure is perceived which in due course is interpreted as being due to the existence of a small number of constituents at a yet more fundamental level. The breakthrough generates much relief, characteristically expressed by some such remark as 'physics has become physics again when it was in danger of becoming chemistry', until the whole process starts once more. The mode of action is spiral rather than cyclic because it is clear that the new level, if not ultimately fundamental, is at least more basic than that which preceded it. This reflects itself in the decreasing size of the objects to which the role of constituent is assigned.

For example, we might start with the idea of chemical elements. Gradually the number of elements known increases. Long before we get to the full 92 their multiplicity has become something of an embarrassment. However, Mendeleef comes along and spots that they can be organized in a way that exhibits certain regularities. His periodic table brings order into the chaos. We are no longer facing a random collection of disparate items but there is structure, with the recurrence of certain properties, so that elements can be grouped into families whose members have similar behaviour. Rutherford then appears and with his 1911 experiment (discussed on p. 94) reveals a tiny central nucleus within the atom. An immense change of scale has taken place. Atoms are approximately a hundred millionth of a centimetre across, nuclei are a few ten million millionths of a centimetre in diameter. Things have grown smaller by a factor of a hundred thousand. Bohr's picture of a solar-system atom with its electrons circling the central nucleus will eventually lead, after many developments in quantum mechanics, to an interpretation of the periodic table. Atoms are now 'large' composite systems, with nuclei and electrons the small basic blocks out of which they are made. Moreover, the transmutation of nuclei in artificially-induced reactions makes it clear that they themselves are aggregates built up from simpler constituents. We have travelled around one coil of the spiral. There is more to come, and it is the task of this book to present it.

Thus described, the investigation of the fundamental structure of matter might sound like the endless circling of a natural-philosophical mulberry bush. Mixed with the beauty and exhilaration of discovery, one might fear that there would be a slight weariness as the ultimate constituents seemed always to elude one, dissolving into their yet smaller components: 'Bigger fleas have lesser fleas . . .'. The apt, inelegant, and often-used metaphor of peeling layers off an onion may perhaps be thought to carry such overtones. In fact the excitement of the chase, and the remarkable surprises which seem so often to lie along the way, are more than adequate to dispel such feelings. Moreover, as we shall see in Chapter VII, the level currently under investigation has certain features which have not been present in the levels which preceded it. The view that one belongs to a uniquely-privileged generation is one that is more frequently found to be due to tricks of intellectual perspective than to the realities of the situation; nevertheless it is just possible that we might have penetrated the centre of the onion.

Curtain up

A suitable point of departure for an account of modern particle physics is provided by the state of the art in 1930. A comparatively small portfolio of fundamental constituents was on offer. The oldest, in terms of human knowledge, was the electron, discovered by J. J. Thomson before the turn of the century. This tiny particle is so small that it requires 10^{27} of them to make up one gram. (10^n is the number corresponding to a one followed by n noughts.) The movement of an electron through matter gives rise to the flow of electricity, so that the charge it carries is a natural unit for the measurement of the charges of microscopic particles. To accord with the sign conventions established for electricity at the beginning of the nineteenth century, the charge of the electron is -1 in these fundamental units. A particle's charge is an example of what we call a *quantum number.* This is simply the magnitude (usually, but not invariably, a whole number in some appropriate units) of an intrinsic property found associated with every specimen of a particular type of particle. All electrons are charged and they all have charge -1 in our units.

Another particle whose existence was known in 1930 was the proton. This also is electrically charged, with charge opposite in sign but equal in magnitude to the charge on the electron, that is $+1$ in our units. The proton differs from the electron, in another way, for it is 1836 times more massive. That is a point to which we shall have to return.

The third, and final, elementary particle known in 1930 was the photon, the corpuscle of light. It had first made itself known in the stirrings of proto-quantum theory in the early 1900s. Planck (in a discovery which he told his son might one day be seen as comparable to those of Newton) had made the extraordinary suggestion that electromagnetic radiation (such as light), instead of being emitted or absorbed in a continuous stream as everyone believed, was in fact emitted or absorbed in packets of energy, or quanta. This is the characteristic signature of quantum theory, the replacement of the continuous by the discrete, *legato* by *staccato.* At first it was not clear whether the discontinuous character applied only to the actual process of emission or absorption or whether the quanta of energy had an abiding significance. Was it like drips from a tap which merge into a continuous volume of fluid as they splash into the basin or were the quanta, the photons, really particles, like shot from a twelve-bore?

The answer came from the photoelectric effect. People had noticed that the way in which certain metals such as selenium became electrically charged when light is shone on them displayed certain puzzling features which seemed inexplicable. In 1905 (the *annus mirabilis* in which he also invented special relativity and explained Brownian motion) Einstein showed that these features were readily understood if light were made up of quanta having energies specified according to Planck's prescription. He pictured the electrons being ejected from the interior of the metal by interaction with a photon. Instead of all electrons being gently agitated by the 'swell' of electromagnetic waves, some were rocketing pheasants hit by the twelve-bore shot which missed the others. Much later, in 1923, Arthur Compton demonstrated directly these billiard-ball-like collisions of electrons and photons.

All this was immensely puzzling. In 1801 Young had demonstrated the wavelike nature of light by exhibiting the phenomenon of interference. Light coming from two slits was found to give a pattern of alternating light and dark on a screen set up a short distance away. Such reinforcements and cancellations are characteristic of waves, as one can see by watching the ripples from two splashes on a pond as they interact with each other. Young's achievement was brilliant, for it disposed of an idea espoused by no less a man than Newton himself. Sir Isaac had believed light to be a stream of tiny corpuscles; now there could be no doubt that it was a wave motion.

This experimental result received theoretical interpretation in the work of Maxwell. His unification of electric and magnetic forces—the most profound discovery of nineteenth century physics—implied that light was waves of electromagnetic radiation, differing only in wavelength from related phenomena such as the radio waves whose existence the theory predicted. If there was one thing in physics which had a sound basis then surely it was the wave character of light. Yet here was the same light behaving in an unambiguously particulate way, as if it knew only Newton and Planck and had never heard of Young or Maxwell. It seemed impossible to reconcile the two insights and equally impossible to do without either of these modes of description if one were to do justice to nature. For a while all people could do was to hold the two in unresolved dialectic tension. Moreover this wave–particle duality proved universal. In 1924, in his Ph.D. dissertation, Prince Louis de Broglie predicted that wave-like properties would also be found to be associated with objects like electrons, which hitherto had been thought of as purely particles. The observation of interference patterns caused by passing a beam of electrons through thin metal foils triumphantly confirmed his ideas.

At the time this must have seemed a substantial challenge to the possibility of finding a consistent rational framework for the description of nature. Yet within a few years a formalism was found that combined the notions of a particle and wave without taint of paradox. This was provided by quantum field theory, the first example of which was constructed by Dirac in 1928. We shall give some account of this formalism in Chapter V.

Character sketches

Certainly the elementary particle scene as revealed in 1930 appeared beguilingly simple: just three fundamental entities, the electron, the proton, and the photon. The picture was to acquire some complications during the 1930s but we can use this over-simple cast-list to illustrate some general features which will persist.

Particles behave as if they were spinning systems. The amount of spin they possess is measured by a quantity called angular momentum, which plays the same role for rotatory motion that ordinary momentum does for motion from point to point. Quantum mechanics provides a natural unit for the measurement of angular momentum, given by Planck's constant, $\hbar$. (Planck called his original constant h, but later developments showed that this quantity divided by 2π was of greater fundamental significance. This is universally written $\hbar$ and pronounced 'aitch slash', 'aitch bar,' or often just 'aitch'.) In terms of this unit the angular momentum can only take integral or half-odd-integral values (that is $0\hbar$, $\frac{1}{2}\hbar$, $1\hbar$, $1\frac{1}{2}\hbar$, $2\hbar$, etc.) Here is another example of the way quantum mechanics makes the hitherto continuous discrete, for in classical mechanics angular momentum can take any value whatsoever. This point is so important that it is worth rubbing it in. Suppose we had a microscopic top and we tried to whip it into motion. Classically (that is to say, in terms of the everyday common sense mechanics of Newton) we could do so as gently as we pleased. The slightest flick of the whip would get it going. Not so quantum mechanically. A flick that could communicate only enough energy to produce $\frac{1}{4}\hbar$ of angular momentum would be nugatory. In fact it is necessary to give the top a whole unit $1\hbar$, since we can show that such rotating systems are only capable of existing with integral values of angular momentum. (The half-odd-integral values arise in a different way, as we shall see in the next paragraph.) The reason why in real life we are blissfully unaware of this remarkable fact is that $\hbar$ is so small a unit. A child's top could easily have an angular momentum in excess of $10^{30}\ \hbar$. The discrete spacing becomes unnoticeable with such large numbers; a row of dots very close together looks like a continuous line.

The particles we have been discussing all have non-zero spin; the electron and proton both have spin $\frac{1}{2}$ (we usually leave out the explicit $\hbar$) and the photon spin 1. Spin is another example of an intrinsic quantum number associated with a particle. The fact that the electron and proton have half-odd-integral spin means that the angular momentum they possess cannot be pictured as due to their being small rotating microscopic systems, for we have noted that motion of this kind can only produce whole-number multiples of $\hbar$. It also turns out that the photon's spin is not to be pictured as arising in this way either, though this is not so obvious a result. We are faced with a paradox.

A point cannot rotate. Our elementary particles are point-like but they succeed in having angular momentum which is characteristic of rotatory motion. This is classically unpicturable, indeed nonsensical, but again it is possible to use quantum field theory to provide a formalism which incorporates this property without any internal contradiction. The moral of the tale is that we need to supplement our intuitive notions drawn from the (classical) everyday world. A careful mathematical investigation of the structure of quantum mechanics, particularly when it is married to relativity theory, provides the necessary enlargement of our imagination.

Another property which particles possess is called *statistics*. This describes the behaviour of aggregates of particles. Do they like being all in the same state of motion or do they prefer to be assorted among many different states of motion? In quantum mechanics there are two contrasting possibilities. Some particles prefer to be in the same state together; you might say that 'the more the merrier' is their motto. They are then said to obey Bose statistics and such particles are called *bosons*. Other particles have the property that you never find more than one of them in a given state of motion. This is the famous Pauli exclusion principle; the presence of one particle in a state excludes all others from this state. These particles are said to obey Fermi statistics and they are called *fermions*. One might illustrate the two types of particles by thinking of people travelling in a not-too crowded train. Englishmen are fermions; they prefer to arrange themselves so that there is not more than one person in each compartment. On the other hand, people of a Mediterranean temperament are bosons; they like to crowd together in jolly groups.

It was noticed that there is a striking correlation between a particle's spin and the statistics it obeys. All particles with integral spin ($0\hbar$, $1\hbar$, etc.) are bosons, all particles with half-odd-integral spin ($\frac{1}{2}\hbar$, $\frac{3}{2}\hbar$, etc.) are fermions. Thus photons are bosons and electrons and protons fermions. This regularity is called the spin-and-statistics theorem. A very interesting fact lies behind it. Quantum mechanics by itself cannot explain the relationship. However, in order to understand elementary particles we have to blend together quantum mechanics (because the particles are small) and special relativity (because they move fast). We are thus in a world doubly removed from that of everyday concepts. The union of these two theories produces an offspring with astonishingly rich and intricate properties. Fifty years after the first consummation we are still investigating the fruits of the marriage. The main discussion of this topic must await Chapter V. For the present we note that Pauli showed that one of the consequences of the union is to enforce the relationship between spin and statistics. In relativistic quantum mechanics integral spin particles must be bosons and half-odd-integral spin particles must be fermions.

This remark raises an intriguing possibility in one's mind. Relativistic effects manifest themselves when one deals with systems with velocities which are an appreciable fraction of the velocity of light (186 000 miles per second). In fact relativistic effects were encountered experimentally well before the spin-and-statistics result was known, so that Einstein provided Pauli with an essential tool. Suppose, however, that velocities comparable to that of light had not been so accessible and that spin-and-statistics had come first. Then it is conceivable that a super-Einstein (for it really would have taxed even that great man's ability) could have predicted relativity from spin-and-statistics. In other words, although special relativity is most characteristically concerned with the properties of fast-moving particles, it also impresses itself on the properties of particles which are slowly moving, or indeed at rest. We shall meet another example of this in the TCP theorem of Chapter III.

The three particles of 1930 were soon joined by a fourth, the neutron, discovered by Chadwick in 1932. It is just a little heavier than the proton, being 1838 times the mass of the electron and, as its name would suggest, electrically neutral rather than charged. It has spin $\frac{1}{2}$ and so is a fermion. It differs from the other particles we have met by being unstable. If a proton or an electron or a photon is left on its own it will last, as far as we know, for ever. Not so the neutron. Trapped inside a nucleus it is stable enough, but on its own it spontaneously decays into other particles after an average time of about 15 minutes (quantum mechanics can only predict the average, individual events will have different times, some shorter, some longer). The matter which made up the neutron has changed into matter in the form of other particles, plus the energy of their motion. In the debris of a neutron decay we always find a proton and an electron. However, people soon realized that there must be another particle as well. This was because without it energy, and momentum, and angular momentum (which we believe are absolutely conserved quantities, that is they never just get lost) did not add up. The energy of the proton and the electron did not add up to the energy present in the neutron, and so on. Since the electric charge did add up correctly (+1 (for the proton)+(−1) (for the electron)=zero(for the neutron)), the third particle must be electrically neutral. It is called the neutrino. For many years the evidence for its existence remained indirect; its rôle was to be the balancing term in reckoning up energy, and momentum, and angular momentum in transactions such as neutron decay. The reason for this elusiveness we shall explain shortly.

The neutrino has an interesting property which it shares with the photon. It has zero mass. A reader with a knowledge of elementary Newtonian mechanics may find this statement strange. If a particle has no mass how can it have energy or momentum? It is not necessary to possess inertia (mass) in order to possess these other qualities? The answer lies in the theory of special relativity. In this theory ordinary particles with non-zero mass actually get more and more massive the faster they move. If they were able to move with the velocity of light (which they can't) they would become infinitely massive. It is not too gross a caricature to say that correspondingly a mass*less* particle can acquire such characteristics of mass as energy and momentum provided it takes care always to travel at the speed of light. The infinity induced by that motion overcomes the zero of its mass to give non-vanishing energy and momentum. (A small number very much augmented becomes a large number, so that it is not a big step to believe that an infinitesimal number infinitely augmented becomes non-vanishing.) This is what the photon does—it *is* light, after all—and so does our new friend the neutrino.

Proto-particle physics

We have now assembled a sufficiently diverse cast to be able to provide a curtain raiser to the drama of particle physics.

Consider first the proton and neutron. They are the building blocks out of which nuclei are made. There must therefore be a force between them which makes them stick together. There are three things which we can say about this nuclear force. First, it must be very much stronger than the forces due to electromagnetism. This is because the electric force between two protons is repulsive and so would blow the nucleus apart were it not overcome by the greater attraction of the nuclear force. In fact the nuclear force proves to be over a hundred times stronger than the electromagnetic forces. The second fact about the nuclear force is that it has a short range; it only makes itself felt over distances of the order of one ten million millionth of a centimetre (a distance which is called one fermi). This again contrasts with electromagnetism. The electric force decreases rather slowly with distance, according to the famous inverse square law, and so makes itself perceptibly felt over long distances. The difference in spatial character between the two forces is responsible for the phenomenon of fission in large nuclei of an element such as uranium. In a large nucleus a proton or a neutron only feels the nuclear force due to its immediate neighbours, the others being too far away to affect it. Therefore the magnitude of the effects due to the cohesive nuclear force increases only in proportion to the number of protons and neutrons, since each has only a fixed number of near-neighbours. The electric repulsion, however, acts between *all* pairs of protons in the nucleus, because of its long range. Thus its effects increase with the square of the number of protons (since if there are Z protons in the nucleus there are $\frac{1}{2}Z(Z-1)$ pairings which can be made between them). Thus the explosive electric forces become more and more significant in relation to the cohesive nuclear forces as nuclei contain more and more protons. This leads eventually to instability and fission.

The third fact about the nuclear force is that it appears to treat neutrons and protons in an exactly similar way. This is called the charge independence of nuclear force, that is the *nuclear* force is the same between two uncharged neutrons, or between two charged protons, or between a neutron and a proton. Again this is obviously in complete contrast to electromagnetic forces which are related to the charges and so differ for protons and neutrons. One of the simplest pieces of evidence for charge independence of the nuclear force is provided by mirror nuclei. These are nuclei in which the numbers of protons and neutrons are interchanged; if one nucleus has Z protons and $(A-Z)$ neutrons (to use a standard notation), the other has $(A-Z)$ protons and Z neutrons. Such pairs (for example, helium-6 with two protons and four neutrons and beryllium-6 with four protons and two neutrons) are found to have very similar properties, with the small differences being of a kind which their different electrical properties are capable of explaining. This fact strongly supports the view that the nuclear force properties of protons and neutrons are identical and this has been confirmed by many other more complicated investigations.

General ideas of great importance for the development of particle physics are contained in the description given above. The first is the fact that there are different types of forces present in nature with different intrinsic strengths and properties. We have so far met two: a strong force (exemplified by the nuclear force) which has large effects over short distances but then dies sharply away; and the electromagnetic force, more than 100 times weaker but extending over long distances. Particles can have very different properties with respect to these forces. For example, as far as electromagnetism is concerned, the proton and neutron could not differ more markedly, whilst they exhibit a surprising similarity in terms of the nuclear force. Of the other particles we have met (the electron, the photon, and the neutrino) none of them is subject to the strong nuclear force and the neutrino, being uncharged, has no electromagnetic interactions either.

Particles which participate in strong interactions are called *hadrons* (from a Greek word for large). The name implies that they are heavy and certainly the proton and neutron fulfill this condition since they are more than 1800 times more massive than the non-strongly-interacting electron. The reason for expecting heaviness to be associated with strongly interacting particles lies in the celebrated mass–energy relation of Einstein,

$$E = mc^2$$

where c is the velocity of light. Particles acquire energy from their interactions (we shall try to explain this in more detail in Chapter V), and this energy translates itself into mass according to Einstein's equation. The stronger the interactions the particles have, the more energy they acquire and the more massive one might expect them to be. This point of view will be subject to some modification as the story unfolds, but it is a good starting point from which to consider the question of particle masses. It enables us to understand why a term meaning 'big' is used to stand for the property of being 'strongly interacting'.

From this point of view the near equality of the masses of the neutron and proton (they differ by about 0.1%) is another indicator of how similarly they participate in the strong interactions from which the bulk of their mass is expected to derive. The difference in their masses is then attributed to the difference in their electromagnetic interactions. At first sight this appears to raise a difficulty. The proton is charged and so would be expected to acquire from electromagnetism extra energy, and hence extra mass. The neutron, having no charge, would not. Yet it is the neutron which is the heavier! If the electromagnetic interactions were determined solely by the charge, or lack of it, this argument would be correct and disastrous. The resolution is rather subtle. Because they have spin the proton and neutron can interact electromagnetically not only as charges but also as if they were tiny magnets. It is presumably this which saves the day, with the neutron's magnetic energy proving sufficiently large to redress the balance. Qualitatively this provides a possible understanding of the proton–neutron mass difference, but it has proved notoriously difficult to make this insight quantitatively successful.

From the point of view of strong nuclear forces the proton and neutron appear to be the same sort of creature. From the point of view of electromagnetic interactions they are clearly different. In 1932 Heisenberg suggested that for strong interactions it would be useful to think of proton and neutron as just two modes of a single object, for which the name *nucleon* was coined. At the time this must have appeared to many to be a formal, even frivolous, suggestion. It would have required unusual percipience to realize that here was the beginning of one of the most fruitful ideas in particle physics. Heisenberg had drawn attention to the existence of the first known example of what we now call a *multiplet.* A multiplet is a collection of particles which have a number of common properties (for example the proton and neutron both have spin $\frac{1}{2}$) and which behave in an identical way with respect to some of the interactions found in nature (as the proton and neutron have identical behaviour with respect to the strong nuclear force). They are differentiated from each other by having contrasting behaviour with respect to other interactions (as the proton and neutron differ electromagnetically). As the number of 'elementary' particles increased (see Chapter IV) their organization into multiplets proved a vital clue in revealing order in what otherwise would have seemed a chaotic plethora. Just as the periodic table with its families of elements helped to elucidate what lay behind atomic physics, so the idea of multiplets will lead us in due course to the structure which lies behind the level of the 'elementary' particles which we are now discussing.

Hadrons are subdivided into two classes. Those with half-odd-integral spin are called *baryons* and those with integral spin are called *mesons.* Thus the proton and neutron are both baryons. We have yet to meet a meson. The names are once more descriptive of the particles' masses, being derived from the Greek for weight and intermediate respectively.

The photon is the next particle to attract our more detailed consideration. Its role is to be the mediator of electromagnetic interactions and the general principle that it illustrates is the way in which forces are pictured as working in the micro world. The old and mysterious Newtonian idea of action-at-a-distance is replaced by the notion that forces are due to the emission and absorption of particles. In the case of electromagnetic interactions the photons are the particles whose exchange gives the force. A force, after all, is simply a mechanism for transferring energy and momentum between the objects on which it acts. If a charged particle emits a photon it gives some of its energy and momentum to that photon and in consequence its own state of motion is correspondingly affected. If another charged particle absorbs the photon it gains for itself the energy and momentum that the photon carried. In this way there has been a net transfer of energy and momentum between the two particles; in a word, a force has acted. All interactions of elementary particles are pictured as occurring in this way. ('What particles then are exchanged to give the strong nuclear force?' an attentive reader may ask. A good question to which we will come shortly. That is where mesons come in.) One can think of it rather like an exchange of broadsides between two ships of the line. The cannon balls are the photons. The ship that fires them recoils and the ship that receives them is knocked off course a little.

The electron and neutrino are particles with spin $\frac{1}{2}$ which do not have strong interactions. Such particles are called *leptons* (from the Greek word for small). It is certainly true that the two examples we have met so far are not very massive. Later we shall meet some more leptons which are really rather heavy so the name is a bit of a misnomer. However, it is now established terminology beyond the possibility of being changed.

The neutrino has neither strong nor electromagnetic interactions. However, it must participate in whatever interaction is responsible for the instability of the neutron, for that is where we first encountered its existence. Neutron decay is an example of a third fundamental class of interactions present in nature. They are called *weak interactions* because they appear a million million times less strong even than the electromagnetic interactions. Weak interactions nevertheless manifest themselves quite widely in nature. They are responsible for many decays of unstable systems; in particular for the so-called β-decays of nuclei which have been studied from the earliest days of radioactivity. It is because it has only weak interactions that the neutrino is so elusive and hard to observe directly. Individual neutrinos can easily penetrate through the earth without any interaction whatsoever—it appears transparent to them. Since we detect particles by their interactions, it is very difficult to pin down neutrinos.

So far we have said nothing about the most famous interaction of all, the gravitational interaction. The reason is that it is so very weak indeed on the elementary particle scale, being only 10^{-37} times the strength of strong interactions (10^{-n} is one divided by (one followed by n noughts)). Therefore we can neglect gravitational effects in elementary particle physics. This may sound surprising. A man would not live long who neglected gravitation in real life. An open window in a tall building would soon see the end of him. The reason why gravitation is so important in the macroscopic world of objects like men and cannon balls is its great cumulative effect. There is only one sign of charge in the interaction with the gravitational field. This means that its effects, though individually small per elementary particle, all add up. There are no cancellations. Everything counts. Compare this with electromagnetic interactions. Here we have plus and minus charges whose effects oppose each other. Since matter is mostly electrically neutral (in ordinary matter there are as many protons as electrons) these electric effects largely cancel out and only in special circumstances (highly charged bodies) are they important on a macroscopic scale. Again, gravitation, like electromagnetism, is a long-range force so that its effects carry over the large distances essential for macroscopic consequences. Here the contrast is with nuclear forces; the latter do not cancel in the way that electromagnetic forces do but their very short range (one fermi) prevents their producing direct macroscopic effects.

Thus while we may forget about gravity for our present purposes, it is nevertheless thought not to differ in character from the other basic forces with which we have to deal. In particular it is believed to be mediated by particle exchange. The hypothetical carrier of gravity is called the *graviton.* In order to play its expected role it would need to be massless and have spin 2. On the basis of general principles we firmly believe in its existence, but the extreme weakness of its interactions makes its direct detection beyond our means. Compared with the graviton the elusive neutrino seems positively assertive of its presence.

We have reached a stage when it may be useful to present a synopsis of the plot so far. In nature we recognize four different types of interaction:

(i) *strong interactions*, exemplified by nuclear forces. They are very strong but act only over short distances of the order of 10^{-13} cm.

(ii) *electromagnetic forces*, only 10^{-2} of the strength of the strong interactions but acting over a long range.

(iii) *weak interactions*, responsible for the neutron instability and other forms of β decay. They are only 10^{-12} of the strength of electromagnetic interactions and are short range in character.

(iv) *gravitational interactions*, long range but only 10^{-37} of the strength of the strong interactions and so negligible in elementary particle physics.

In the everyday world of macroscopic objects (men and cannon balls) the effect of strong interactions is to hold nuclei together and so make possible the variety of chemical elements. The electromagnetic forces determine the structure of atoms and the long-range forces which hold liquids and solids together. Only gravitation provides a fundamental force which, because of its cumulative character, is *directly* perceptible macroscopically in all ordinary circumstances.

In addition to dividing fundamental forces into classes we have identified some generic types of particles:

(i) *hadrons*, which have strong interactions. The idea of multiplets, illustrated by the (neutron–proton) nucleon doublet, will prove an important clue to later structural developments. Hadrons divide into baryons (with spin $\frac{1}{2}$, $\frac{3}{2}$, etc.) and mesons (with spin 0, 1, etc.).

(ii) *leptons*, which have spin $\frac{1}{2}$ but do not have strong interactions. The neutrino catches our particular attention as being massless and having only weak interactions.

(iii) *the photon*, which as the object exchanged in electromagnetic interactions illustrates the important idea that force in elementary particle physics will be transmitted by particles which act as carriers of energy and momentum.

Further developments

Two important theoretical developments in the 1930s led in turn to further experimental discoveries of great significance. The first of these developments was the concept of matter and antimatter.

In 1928 Dirac had invented a relativistic wave equation which described the behaviour of spin-$\frac{1}{2}$ particles such as the electron or proton. It was a brilliant piece of work—one of the great triumphs of theoretical physics in this century—and had all the elegance which makes one feel instinctively that it 'must be right'. Moreover, the equation also scored a notable success in predicting correctly something called the gyromagnetic ratio of the electron.

We have said that particles of spin $\frac{1}{2}$ can interact with the electromagnetic field not only through their electric charge but also in a way that corresponds to their acting like tiny magnets. Now a magnet has a north pole and a south pole and so has a direction associated with it, corresponding to the line joining the two poles. If an electron behaves like a tiny magnet, what decides the direction in which this magnet points? The answer is that it points along the direction of the axis of rotation associated with the electron's spin angular momentum. If one were allowed to discuss all this classically a simple picture would emerge. The spin would be due to the material in the electron rotating; this material must be electrically charged; moving charge is just what we call a current; as long ago as Faraday's time we learnt that currents produce magnetic effects. Thus not only would this provide a qualitative understanding of why spinning particles behave like magnets but it would also furnish a quantitative description. The faster the charge rotated, the bigger the current would be, and so the bigger the magnetic field. Thus this classical picture would imply a proportionality relating the magnetic properties (magnetic moment) to the rotation (angular momentum). The constant of proportionality is called the gyromagnetic ratio. In actual fact measurement shows that the electron has a gyromagnetic ratio which is just twice that expected on the basis of the classical argument we have outlined. This should occasion no surprise. After all we have already argued that the intrinsic spin of elementary (point) particles *cannot* be thought of as corresponding to the rotation of a small extended system! The spin is classically unpicturable and the argument given above has the bottom knocked out of it in consequence. However, that seems to imply that we have no basis for predicting what the gyromagnetic ratio should be, a scientifically unsatisfactory agnosticism. Dirac showed that his equation overcame this difficulty. It predicted the right value for the electron's gyromagnetic ratio. This was a great triumph, the more so because the result emerged as a quite unexpected bonus from the apparently unrelated argument which had led Dirac to postulate his equation in the first place.

There was, however, a fly in the ointment. Excellent though the Dirac equation proved to be, it had one apparently disastrous feature. The equation admitted a variety of solutions. Some were perfectly suitable for the description of electrons, but others were not, for they gave negative values for the energy. This was, of course, quite unacceptable. Moreover, quantum mechanics would allow transitions from the acceptable positive energy states to these unwanted negative energy states. Thus the scandal could not be ignored. Dirac brooded on the problem and at length came up with a daring solution. There was only one thing which could stop transitions to the negative energy states—the exclusion principle (p. 7). Electrons are fermions and so there can never be more than one in any given state. Suppose that the ordinary state of the world is one in which all the negative-energy electron states *are already filled up*. Then effectively the exclusion principle blocks them off—it is as if they weren't there after all. The problem is solved! Thus Dirac proposed a picture of the world in which we live as permeated by a 'sea' of negative-energy electrons. But doesn't that mean an awful lot of charge and (negative) energy lying around in these electrons? Well yes, but in fact it is only *changes* in quantities like charge and energy which count; the zero from which you measure is arbitrary, so that doesn't matter. The reader who feels uneasy can be assured that in Chapter V we shall encounter quantum field theory which provides another way of thinking of the problem and retains the desirable consequences of Dirac's idea without adhering to the bizarre literal interpretation of a negative-energy electron sea.

Dirac saw that his idea not only solved the negative-energy problem but also had an astonishing further consequence. Suppose that a very energetic photon is absorbed by one of the negative-energy electrons in the sea with the consequence that, on adding to itself the photon energy, it becomes a positive-energy electron. The photon will have to be pretty energetic, for it will have to cancel out the negative energy of the sea electron (which is at least minus the rest mass energy, that is at least $-mc^2$) and provide the positive energy of the new electron into which it is transformed (which has energy at least $+mc^2$). Therefore the photon itself must have an energy not less than $2mc^2$. As a result of such a process two new objects have been created. One is the new positive-energy electron, which is simply an honest-to-goodness electron with nothing more to be said about it. The other object is due to the fact that there is now a 'hole' in the sea. One of the negative-energy-state electrons has been removed. Now two minuses make a plus and the absence of a negative-energy object will correspond to something which behaves like the presence of a positive-energy object. What is this new 'thing' which has appeared? Because it corresponds to the absence of a negatively charged electron it will appear to be positively charged. Thus holes in the sea will behave like some new positively charged brothers of the electron.

At this point a curious twist occurs in the story. It is clear from the argument that the new positively charged object must be very similar to the electron and must in fact have the same mass. No such particle was then known and even Dirac did not have quite sufficient courage of his convictions to predict the existence of a totally new form of matter. For a while he ignored the mass problem and toyed with the idea of identifying this new object with the only positively charged particle to hand, the proton. Oppenheimer trenchantly pointed out that this would not do. In the end there seemed nothing left but to see if there really was a positively charged brother of the electron with the same mass. Of course, there was. The positron, as it is called, was soon discovered in cosmic rays. In fact it was found that positrons had appeared for many years in cloud-chamber photographs and had received some comment as anomalous events whose interpretation was unclear! It is very easy only to see what one is told to look for.

The relation of electron and positron illustrates a duality to be found in all of nature—every particle has associated with it an antiparticle of the same mass and opposite sign of charge. Thus the proton has a negatively charged brother, the antiproton, which was not discovered experimentally till 1955. The historical way in which we have introduced the phenomenon, via Dirac's struggle with the negative-energy solutions of his equation, might seem to suggest that particle–antiparticle duality is a special feature of fermions. Not so, for matter-antimatter is a universal bifurcation in nature, modified only by the fact that some particles can span the divide by being their own antiparticles. Clearly this is only possible for a neutral particle, since charge has to change sign between particle and antiparticle. The photon is an example of a particle which is its own antiparticle. On the other hand, not all neutral particles behave in this way. The antineutrino is distinct from the neutrino.

A sufficiently energetic photon can always create a charged particle–antiparticle pair, in the way that we considered for the electron–positron. Conversely a particle and antiparticle can combine (annihilate) to release their mass–energy either electromagnetically or in some other way.

The laws of nature are almost (but not quite, see Chapter III) symmetric between particle and antiparticle. Yet around us particles predominate. Electrons and protons are the stuff of which our world is made whilst positrons and antiprotons are rare, found only in cosmic rays or created in laboratories. Why is this so? Are there other parts of the universe where antimatter predominates over matter or is it just that at the 'big bang' things started in a material rather than anti-material way? We don't know for sure the answer to these questions.

The second fruitful development of the 1930s brings mesons on to the stage. Again the story is something of a triumph for theory since it starts in the study rather than the laboratory. The hero of this interlude is the Japanese physicist Hidekei Yukawa. We have spoken of photons acting as the carriers of electromagnetic forces. Yukawa asked the question 'What particle acts as the carrier of the nuclear force?' He recognized that the spatial characteristics of the force are directly related to the mass of the particle which plays this role. The electromagnetic force is long range because the photon is a massless particle. Conversely the short range of the nuclear force implies that the particle which mediates it must have non-zero mass. In fact Yukawa was able to predict what its mass should be. The range r_0 of the force is related to the mass m of the particle by the formula:

$$r_0 = \frac{\hbar}{mc}.$$

The other two quantities in this equation are Planck's constant $\hbar$ (characteristic of quantum theory) and the velocity of light c (characteristic of relativity). To give a range of about 1 fermi (10^{-13} cm) the particle must have a mass two or three hundred times that of the electron. It was for this reason that the word *meson* (originally, mesotron) was coined, for the mass is intermediate between that of leptons and baryons. (As with all these generic names, examples of the species were found much later which make the terminology ill judged.)

Soon after Yukawa made his suggestion of the meson as the mediator of the nuclear force, a charged particle was found in cosmic rays which was about 200 times the mass of the electron. It appeared a triumphant vindication of the theory. Before the question could be pursued in much detail the war intervened and the minds of nuclear physicists were directed to other problems. After the war, when the pursuit of pure science returned, a curious fact emerged. If this new particle were playing its expected role then it would, of course, be a hadron and should interact strongly with nuclei. To everyone's surprise this did not happen. In passing through matter the putative meson exhibited electromagnetic interactions only. The resolution proved most satisfactory. A new phenomenon was revealed in such a way that no one lost face.

In fact there were two sorts of new particle. One was a hadron called the π-meson (pi-meson or pion) which played the role Yukawa had assigned to it. When it was eventually identified it proved to have a mass some 270 times that of the electron. It was however unstable, rather like the neutron, and it decayed into the particle which had already been found (now called the μ-meson (mu-meson or muon) and a neutrino:

$$\pi^+ \rightarrow \mu^+ + \nu.$$

We are here using standard notation; particles are represented by the corresponding symbols with $\pm$ superscripts indicating their electric charges. An arrow shows the way the reaction goes. The π-meson—pion for short—is a good deal quicker about its decay than the neutron. The latter lasts on average for 15 minutes, the pion only for about 10^{-8} seconds. So transient an existence might be thought scarcely to justify the dignity of being called a particle. However, the typical time involved in strong interactions—that is to say the average time between emission and absorption of a pion—is very much shorter, 10^{-23} seconds in fact. Thus from the point of view of strong interactions the pion is as stable as need be. It is of course the weak interaction which brings about its ultimate downfall, via the process shown.

The pion has spin zero and occurs in three charged states, the

π^+; its antiparticle the π^-, which decays

$$\pi^- \rightarrow \mu^- + \bar{\nu},$$

in analogy with π^+, the bar indicating that it is an *anti*-neutrino in the final state; and a neutral π^0 (which is its own antiparticle). The π^0 has an electromagnetic (rather than weak) decay, into two photons,

$$\pi^0 \rightarrow \gamma + \gamma,$$

the symbol γ (gamma) denoting a photon. Because the electromagnetic interactions are stronger than weak, π^0-decay goes more rapidly than π^+-decays and the π^0 has a lifetime of only 10^{-15} seconds. Again of course, this is very long compared with the typical strong-interaction time of 10^{-23} seconds.

Nuclear forces are due to the emission and absorption of pions. If a proton emits a π^+ it has to turn into a neutron to conserve charge. It can also emit a π^0 and stay a proton, though with energy and momentum reduced by the amount corresponding to that carried off by the pion. A proton cannot, however, emit a π^- since there is no doubly-charged nucleon for it to turn into, as it would need to do in order to conserve charge. On the other hand, a neutron can emit a π^- and become a proton in the process, or a π^0 and stay as it is, but it cannot emit a π^+ for there is no negatively charged nucleon for it to turn into in the effort to conserve charge. Similarly protons can only absorb π^-s and π^0s and neutrons can only absorb π^+s and π^0s. Despite all these rather complicated constraints on what can happen, which depend on the charge of the nucleon involved, the charge independence of nuclear forces asserts that the resulting total force between two nucleons does not depend upon their charge states. If this is true, some quite clever relationships must exist between the various emission and absorption processes we have discussed. There is in fact, as Nicholas Kemmer showed, a unique solution to the question of what these relationships must be. This will prove a fundamental clue to particle structure when we return to the discussion of these matters in Chapter IV.

For the present, however, let us return to the μ-meson, or muon for short. It was found to have spin $\frac{1}{2}$ so that it is not a meson at all—the name is due to the historical accident of its original misidentification. Rather, since it has no strong interaction, it is a lepton like the electron. In its turn it is unstable, living for about 10^{-6} seconds and then undergoing a weak decay into an electron (or positron), neutrino, and antineutrino,

$$\mu^{\pm} \rightarrow e^{\pm} + \nu + \bar{\nu}.$$

We have here written two equations for the price of one, using $\pm$ to denote the two charge states of muon and electron–positron. The reason the muon decays into an electron rather than vice versa is simply a question of mass. The muon is heavier and can dispose of its surplus mass as kinetic energy (energy of motion) distributed among the final particles of its decay. The electron is forbidden by energy conservation to turn itself into the heavier muon. In fact, when their properties are examined in detail, the electron and muon appear absolutely identical, mass only excepted. Two important remarks follow from this observation.

The first is that here we encounter for the first time a mysterious multiplicity in nature. The muon is simply a heavy electron; it reveals no new insight into the structure of matter other than the existence of this duplication. 'Who ordered the muon?' In an economic world one could suppose we could do without it. For a long time the muon was the only example of this sort of superfluity. However, we shall find (in Chapter VIII) that the most recent developments of particle physics have shown that such prodigality is a widespread feature of the natural scene.

The second remark instigated by contemplating the muon is that the simple idea, explained earlier in this chapter, that particle masses are simply the Einsteinian translation ($E = mc^2$) of their interaction energies cannot be adequate to explain the world. In terms of basic *interactions* the electron and muson are identical yet one is 206 times more massive than the other. A story is told of the celebrated American theoretical physicist Richard Feynman. He had been brooding on the origin of the particle masses and at first thought how neatly they seemed to reflect the strengths of their interactions. However, when a friend found him later he was hopping with frustration, crying out 'Godamn the μ-meson.'

The story so far

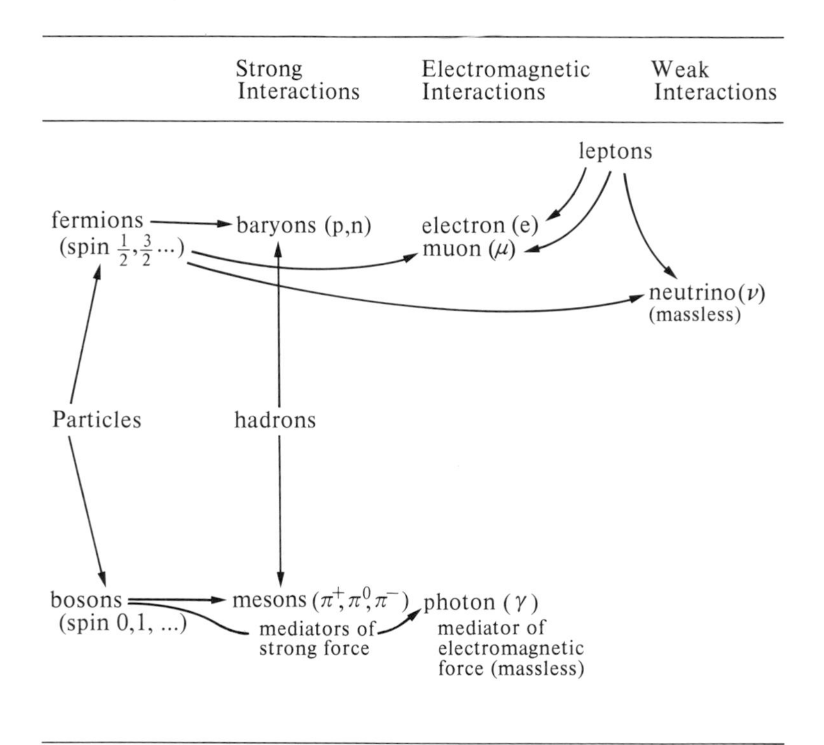

For each particle its strongest interaction is shown. In general it has all the interactions to the right of its entry, so that baryons, for example, have electromagnetic and weak interactions in addition to their strong interaction.

II
Stage Machinery

It is clearly no easy task to investigate objects which are no bigger than 10^{-13} cm across. You cannot pick them up in your fingers nor even examine them under an electron microscope. Rather it has to be done by banging them together and seeing what happens. In consequence almost all experiments in particle physics are scattering experiments. That is to say, they involve hurling a fast particle at another particle and observing the outcome. The projectile particle may bounce off the target or may break it up. From the patient analysis of the results of many collisions it is possible to build up a picture of the nature of the colliding objects. It is rather as though we were trying to find out what an unknown object was by seeing how ping-pong balls bounced off it, together with the occasional cannon ball that knocked it to smithereens. Not the simplest method of investigation, perhaps, but when it comes to elementary particles it is the best we have got.

Two pieces of apparatus are indispensible for such experiments. One is some form of accelerator to produce the fast projectile particles. The other is a device for detecting the results of the collision. The history of particle physics is, from one point of view, the story of the development of ever more powerful accelerators and ever more subtle detectors. The men who invented new techniques for these purposes rightly figure largely in the lists of Nobel prize winners. Without the tools they provided the job could not have been done.

To make particles move fast you have to find some way of feeding them energy. (High-energy physics is an alternative name for elementary particle physics.) The simplest way to do so is to use an electromagnetic field as the conveyor, so that the energy is transferred to the particles from radiofrequency (r.f.) power supplies. A consequence is that only charged particles can be

The CERN site. The ring of the PS can be seen on the right and the ISR ring is visible left of centre. The large broken circle indicates the position of the underground ring of the SPS (*photograph* – CERN).

accelerated. The machines come in two shapes, straight or circular. In the straight *linear accelerators* the particles travel from one end to the other getting faster and faster, rather like a stone falling under gravity, except that here the force is electric and in consequence it is convenient for the motion to be horizontal. Soon the particles are moving with velocities which are sizeable fractions of the velocity of light. This means they cover a lot of ground in a short time, so that for high energies the machines have to be inordinately long. At Stanford there is a linear accelerator for electrons which is two miles in length. Precision engineering on this scale makes enormous demands on advanced technology and on the scientific budget.

For most purposes, including all machines currently operating to accelerate protons, it is only practical to use a linear accelerator initially in order to get things on the move. The particles are then fed into a circular machine, called a *synchroton*, for their full acceleration. Here they are bent round in a circle by powerful magnetic fields. They can then make many circuits, getting a boost from the r.f. power each time they go round. Part of this energy makes up for energy lost by radiation as the charged particles are bent by the magnetic fields (the effect is called synchroton radiation and is only important for electrons) and part serves to make the particles move faster. It then requires a stronger magnetic field to hold them to the same circle so this has to be increased. The device gets its name from the necessity to synchronize this requisite increase in magnetic field with the increase in the particles' energy. In the old-fashioned *cyclotrons* this was not done and the particles spiralled outwards as their energy increased in the fixed magnetic field. In consequence the cyclotron needed an enormous and expensive disc-shaped magnet. In the synchroton the magnets form a ring along the chosen circle of motion with a consequent substantial saving in steel and money.

The two largest proton synchrotons are located at Fermilab, outside Chicago, and at Geneva. The latter machine is at CERN, the international European collaborative laboratory financed by twelve member states, including the United Kingdom. The CERN accelerator is called the Super Proton Synchroton or, acronymically, SPS. It forms a ring over 2 km in diameter, contained in a tunnel on average 40 m below ground, and

Part of the ring of the SPS accelerator (*photograph* – CERN).

straddling the Swiss–French border. 10^{13} protons are injected into the machine and while they are being accelerated these hardened travellers cross and recross the border at a rate of 100 000 times a second. The ring contains 1000 electromagnets, each positioned to an accuracy of a tenth of a millimetre. It is not very surprising to learn therefore that the SPS took nearly seven years to build at a cost of 1150 million Swiss francs. Somewhat over 16% of the money was provided by the British tax payer, say £60 million ($120 million) in round terms at present rates of exchange. One of the reasons for trying to describe high energy physics is that people are entitled to know what they are paying for.

The fast particles from the synchroton can be used in two ways after extraction from the machine. One is to make them impinge on a fixed target, a lump of some suitable material. For example, if the target is made of liquid hydrogen we can study proton–proton interactions as our fast protons interact with the protons which are the nuclei of the hydrogen atoms in the target. These reactions can be studied for their own sake or they can act as generators of secondary beams of particles whose interactions can in turn be investigated as they are made to strike a further target. For example, protons passing through matter can undergo what is called a charge-exchange reaction in which they lose their charge to the matter and turn themselves into neutrons. In this way a secondary beam of fast neutrons would be produced which could not be made in an accelerator directly because of the neutron's zero electric charge. Such fixed target experiments have been the staple diet of experimental particle physics for many years.

In recent years a second way of using beams of high energy particles has been developed. The particles can be stored in a device called an *intersecting storage ring* to form two circulating beams of particles. Once sufficient particles have been accumulated, the two beams can be made to collide and interact. In this way we can attain much higher energies than are available by simply letting a single beam hit a fixed target. This is because in the latter case only the projectile is moving and so it has to provide all the high energy, the particle in the target not having much more than its rest mass energy to offer. In a storage ring, on the other hand, both particles contribute their high energy to

One of the intersection regions of the ISR where the proton beams are made to collide (*photograph* – CERN).

the total energy present in the interaction. The largest storage ring containing protons is the ISR at CERN. It gives access to energy ranges more than double those accessible from the SPS in its fixed target mode, despite the fact that the protons in the ISR only come from a synchroton (called the PS) which operates at less than a tenth of the SPS energy. Plans are afoot to use the SPS as a collider and when this happens a new energy range will be opened up ten times bigger than that available at the ISR.

This advantage of high energy in a storage ring is purchased at a price. The beams are much less dense than the matter contained in a fixed target. This means that the number of reactions occurring is smaller and the means for investigating them have to be correspondingly more delicate. In particular, only primary interactions can be studied and there are no secondary beams available involving other particles. Nevertheless intersecting storage rings have proved a particularly powerful addition to the particle physicists' range of machinery. At the present time there is special interest in machines which collide an electron beam with a positron beam. The resulting annihilation creates a blob of pure energy which can materialize in an impartial way into many interesting states of matter. (More of this in Chapter VIII.)

The collision having taken place, the next necessity is for the second type of apparatus, some form of detecting system to see what has happened. We must somehow get the particles which fly off to make their presence felt. By far the simplest way for them to do so is through their electromagnetic interactions in matter but it is only on the charged particles that tabs can be kept directly in this way. The behaviour of neutral particles will have, as far as possible, to be inferred from other considerations. For example, if when all charged particles are accounted for there is still some missing energy and momentum in the final state, we know it must have been carried off by one or more neutral particles.

One of the earliest particle detectors was the cloud chamber, invented by the Cambridge meteorologist, C. T. R. Wilson. His real interest was in the formation of Scotch mist but, in an early example of technological spin-off, he also provided a way of making particle tracks visible. A vessel contains super-cooled water vapour which is about to condense. A fast charged particle moving through the vapour ionizes some of the atoms that it

BEBC (the Big European Bubble Chamber) (*photograph* – CERN).

passes and they form sites at which tiny drops of water can condense. The resulting chain of droplets makes the track of the particle apparent. Cloud chambers are no longer used in high energy physics; instead they have been replaced by bubble chambers. These employ the reverse process, for they contain a liquid about to boil, often a cooled-down gas such as liquid hydrogen or helium. The charged particles then leave a trail of tiny bubbles behind them. The inventor of the device, Don Glaser, is said to have got the idea whilst watching the bubbles rise in a glass of beer. Bubble chambers combine target and detector, for the interactions they are used to study take place in the chamber itself, between the incident projectiles and the particles making up the liquid which fills it. Bubble chambers can also exhibit the presence of photons in an interaction, for in the rather dense liquid of the chamber a high energy photon can turn itself into a electron–positron pair which then leaves identifiable tracks behind it.

Many other forms of detector exist which are based on clever electronic monitoring of the passage of charged particles. The presence of the particles is manifested either by flashes of light that they produce (scintillation or Čerenkov counters) or by electrical disturbances between parallel planes of conductors (spark chambers, proportional wire chambers). Some of these detectors are very large. One used in neutrino experiments at CERN contains 1400 tonnes of steel. Such objects are not cheap to build.

The great merit of electronic devices is that they can handle very large numbers of interactions occurring in short times. However, they are less good than bubble chambers at extracting the maximum information available about a given event. Electronic detectors are mostly used to look for some rather specific process which they are designed to pick out from the welter of events taking place. This they can do with great accuracy and can rapidly accumulate many instances for study. Bubble chambers are more plodding but give a more impartial view of what is going on—'slow but sure' one might say. The two types of detector play complementary roles and experiments using both are necessary. Recent developments in technique sometimes make use of a hybrid device in which electronic selection is made of interesting-looking events and an associated bubble chamber is used to study them in more detail.

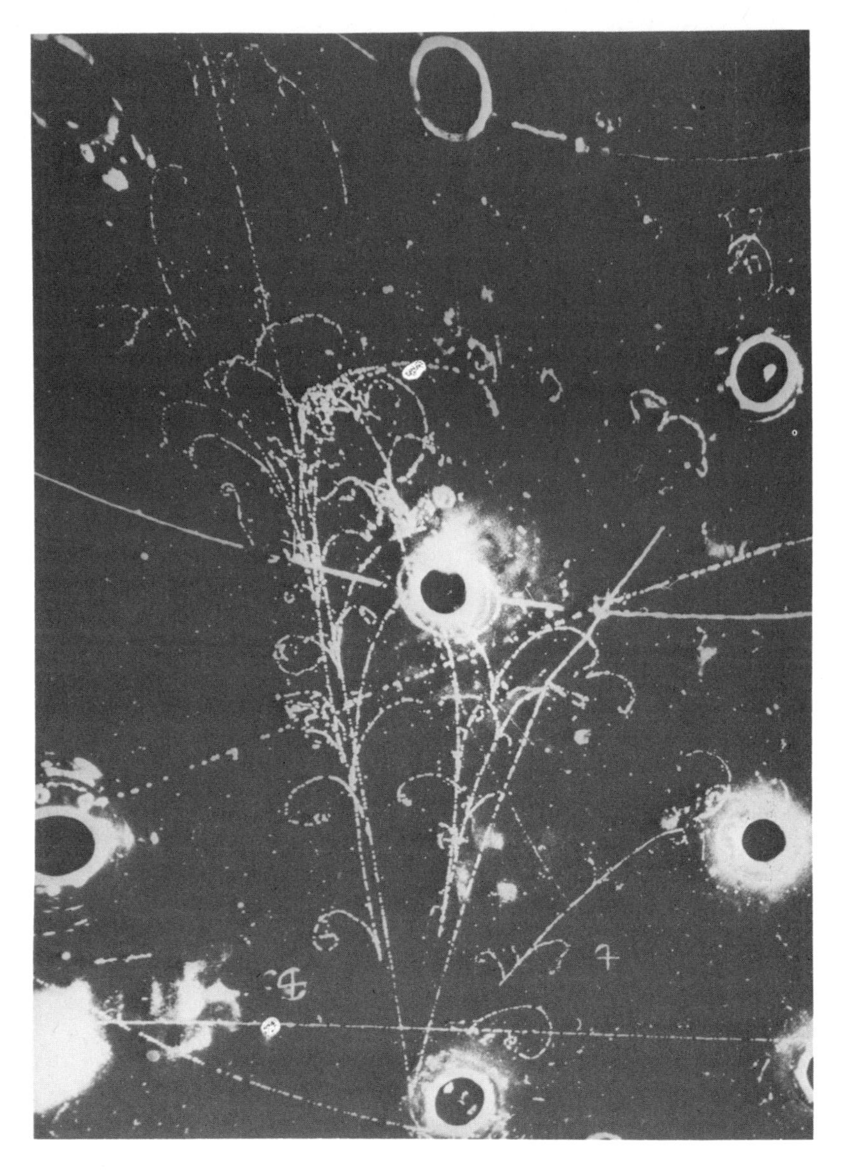

A charged current weak interaction (Chapter VIII) in a CERN bubble chamber called Gargamelle. An invisible neutrino enters at the bottom and causes the interaction originating in the 'V' at the foot of the picture. A muon track forms the left arm of the 'V' and a proton track the right arm. Invisible π^0s are also created in the interaction and it is possible to see subsidiary 'V''s which are the electron–positron pairs created by the photons which are themselves the result of the decays of these π^0s. A magnetic field is applied to the chamber which makes negatively-charged particles bend to the left and positively-charged particles bend to the right (*photograph – CERN*).

The 1400 tonne electronic target-calorimeter used at CERN to study neutrino interactions (*photograph*–CERN).

III

Setting the Scene

Conservation laws and symmetries

It is obvious that our experience of the physical world varies from place to place. Here am I warm and dry in my holiday cottage whilst a few hours ago I was weather beaten and rain drenched on top of the fell. So it is also for elementary particles. The protons in a synchroton are subject to strong electric and magnetic fields but when they are released and impinge on the target it is their strong interactions with its constituent nuclei which predominate. However, these variations of experience are variations of circumstance only. In each case, if a detailed analysis were made, the fundamental laws of physics which were operating on these differing occasions would be found to be the same. It is simply that variations in the environment produce variations in the consequences which flow from these laws. We can appropriate to science a slogan from a different discipline and say that what we have to deal with is what is true at all times and all places. ('Quod semper, quod ubique', as was said of theology.) Were it not so, science with its celebrated repeatability of experiments would not be possible.

We express this fact by saying that the fundamental laws of nature are translation invariant. A translation, in this sense, is simply transference from point A to point B. It can be spatial translation, from here to there, or a temporal translation from now to an earlier or later time. Wherever and whenever B is in relation to A, we shall find the same laws of physics there as we did at A. All that may strike you as interesting but, perhaps, a shade philosophical. It is a surprising fact, though, that from this property that the laws of physics are translation invariant, we can deduce some powerful consequences. Spatial translation invariance implies the conservation of momentum and temporal translation invariance implies that energy is conserved! That is to say, when two particles collide the basic reason why the momentum lost by one is just equal to that gained by the other is that physics here is the same as physics there. The demonstration that this is so requires a bit of mathematics. In a book of this character I shall have to ask my readers to trust me that it is the case. These results are striking examples of a fundamental relationship which is basic to an understanding of particle physics. This relationship is the connection between *symmetry principles* and *conservation laws.*

We must brood on this remarkable fact for a while. A symmetry is an expression of an equivalence between things. A circle is a very symmetrical geometrical figure because each point on its circumference is related to the other points in an identical way, so each point is equivalent, or as good as, any other. An equilateral triangle (that is one with equal sides) is obviously less symmetrical than a circle because its vertices clearly have a different character from points along its sides. Nevertheless it retains some symmetry because each side is equivalent to either of the other two sides. That symmetry would be lost if the triangle were no longer equilateral but had sides of differing lengths. A conservation law, on the other hand, is the statement that there is some quantity around which is preserved in transactions. Provided we exclude the activities of counterfeiters and the mint, coinage is a conserved quantity. It may transfer from my pocket to yours but it does not get lost. The property of symmetry and the property of conservation just sound so different in character that one would have to be very clever first to suspect a connection between them.

Translation invariance is a symmetry principle. It implies that any point of space or moment of time is the same as any other as far as the fundamental laws of physics are concerned. These points of space, or moments of time, are symmetrically related to one another, there is no intrinsic distinction between them. Of course such an idea has not always been part of man's view of nature. In Aristotelian cosmology sub-lunary matter and the heavenly spheres beyond were quite distinct in character.

A conservation law, we have said, refers to some quantity which can never get lost; the sum total of it around is always the same. The conservation of energy and of momentum give prime examples of such laws. The conservation of electric charge is another example we have encountered.

Symmetry principles imply conservation laws. It is important to grasp this fact even if the proof of it eludes one. Conversely, when we find a conservation law operating it is natural to ask what symmetry principle is behind it. The alert reader will ask 'What symmetry principle then is behind the conservation of charge?' Unfortunately the answer is a technical one, only intelligible to those who have some acquaintance with the theory of Maxwell's equations for electromagnetism. For the record, conservation of charge follows from gauge invariance.

The laws of the nature are symmetric in another way. If I face one way and do some experiments I shall find the same fundamental laws as if I were facing another direction. We can characterize this by saying that there are no fundamental directions in nature. (If there were, things would behave differently along them to the way they did at right angles to them, so the way I faced would matter.) Again it is very important to recognize that we are talking about the *fundamental* character of laws. In many situations there will be special directions defined by the environment which is being experienced. For men on a mountain up hill or down hill will be special; for protons the direction of some electric field applied to them equally will be so. However, that it not true of the fundamental laws which control the way the environment produces its influence. We can summarize this by saying that the laws of physics are invariant under rotations; they do not change when I rotate myself to face now this way, now that. There must be a corresponding conservation law. Just as momentum is connected with moving from point to point (translations), so angular momentum is connected with rotatory motion. Therefore it is the conservation of angular momentum which follows from the invariance of the laws of physics under rotations.

Reflections

Until 1956 people believed that the laws of physics were also invariant under reflections. This would mean that if you studied a physical system solely by means of watching its reflection in a mirror you would find the same laws of physics as you would discern by observing the system directly. Of course this would be rather an odd way of actually conducting an experiment. The description is simply a pictorial representation of what is implied by requiring physics to be unchanged by reflections. Important consequences flow from imposing this constraint on the structure of a theory. For example, it would mean that there could be no fundamental preference in nature for right-handedness over left-handedness. This is because one turns into the other under reflection, so that in the mirror the preference, were it to exist, would appear the other way round. Again we are talking about the basic laws of physics rather than effects that result from the way things are constructed. Clearly if we look at human beings there is a preference for right-handedness over left-handedness, since most of us are right-handed. However, this is explained by the way our bodies are put together rather than by its being a consequence of fundamental laws. Most corkscrews are right-handed, but that is due to the way they are made rather than to an intrinsic right-handedness in the original lump of iron.

It seemed intuitively obvious to people that nature must be even-handed. As far as strong and electromagnetic interactions were concerned this was correct, but in applying the principle to weak interactions greater and greater difficulties developed (we will describe what they were shortly). Eventually two American-Chinese physicists, Lee and Yang, were bold enough to suggest that maybe the weak interactions showed a preference for the left-hand. This brought the scornful retort from Wolfgang Pauli that he could not believe 'that God was a weak left-hander'. However, in such matters the last laugh is with the experimentalists and they were encouraged actually to look to see if weak interactions had a handedness. Such experiments had been possible for a long time (there are all sorts of hard-luck stories about people who had thought of doing them but had not quite got round to it). But no one had bothered because they thought they knew the answer. Now when it was tried it was found that indeed weak interactions are not reflection invariant.

One of these experiments is schematically illustrated in the figure. A lump of matter containing some cobalt-60 is placed in the centre of a coil of wire carrying a current. The Co^{60} nucleus is β-decay unstable, turning into nickel-60 and emitting an electron and antineutrino. This decay is a weak interaction effect.

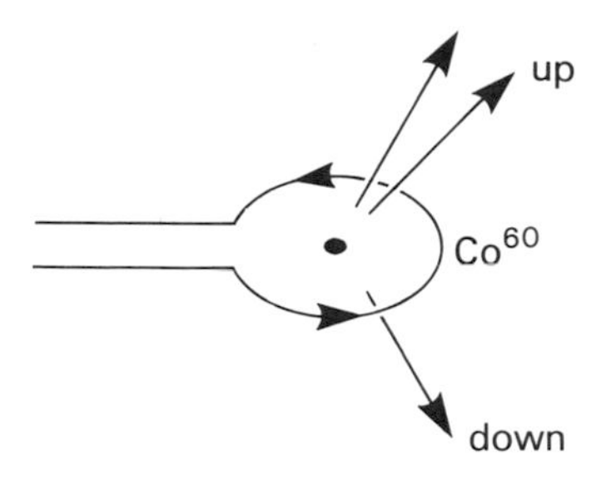

A current is passed through the coil and the emitted electron observed. As individual nuclei decay the electrons come off in various directions, but it is found that more are emitted in the up-direction from the plane of the coil than the down-direction. Up and down here have nothing to do with gravity (it is too feeble), but are defined by the direction of the current in the coil. This fact is confirmed when it is found that reversing the current in the coil reverses the direction in which more electrons are emitted. But how can we define up and down from the direction of current flowing round a coil? Only by using something like a right-hand rule which defines the up-direction as that in which a right-handed corkscrew points if the direction of its rotation is the same as the direction of circulation of the current. But, of course that is just the sort of handedness which is only possible if the weak interactions are not reflection invariant. In a mirror 'up' and 'down' defined this way would interchange.

The reader who has done some elementary school physics may feel a little uneasy at this point. Wasn't there some-one-or-other's right-hand rule for determining the direction of the magnetic field from the way the current flowed round a coil? There was. Did that not then already imply a breakdown of even-handedness? The answer to this is no, but the reason is a little subtle. (It is worth trying to explain the point but those whose minds are untroubled may wish to skip on to the next paragraph.) We are really discussing *vectors*, that is quantities which have both magnitude and direction and which therefore can be represented by arrows. The electron's momentum is a vector, so is the magnetic field due to the coil. Now there are in the world two different sorts of vector, those whose direction corresponds to a line *along* which something goes and those whose direction corresponds to a line *around* which something goes. Technically these two types are called polar and axial vectors, respectively. The electron's momentum is a polar vector; it points along the direction in which the electron is travelling. The magnetic field is an axial vector, since its direction is determined by the axis of the coil around which the current flows. Axial vectors need a right-hand rule in order to specify which way round to go. Polar vectors have no such need, 'along' is unambigously in the direction of the arrow. This difference means that the two sorts of vectors have different properties under reflections and so if the laws of physics are to look the same in a mirror those laws must only link together vectors of the same type. It is perfectly all right to compare my right hand and a right-handed corkscrew. In a mirror they will still be similar, even if they now both appear to be left-handed. What we must not do, if we are to preserve reflection invariance, is to compare something possessing an intrinsic handness with something that does not. In the jargon, we must not correlate an axial and a polar vector. The right-hand rule for magnetic fields gives no trouble because it links one axial vector (defining the direction of the axis of the coil) with another (the magnetic field). On the other hand, the cobalt-60 (Co^{60}) experiment gives a breakdown of reflection invariance because it correlates an axial vector with a polar vector. The way it does so is worth spelling out. How does the Co^{60} know in what direction the current circulates in the coil? The information is carried to it by the magnetic field of the coil. In Chapter I we saw that spinning systems can behave like tiny magnets which point along the axis of rotation. The orientation of these magnets, and consequently the direction of spin of the system, are affected by the magnetic field which they experience. For Co^{60} this has the consequence that the nucleus's spin tends to align antiparallel (that is, in the opposite direction) to the magnetic field. (This alignment is made more effective by doing the experiment at very low temperatures, a technical point of great experimental importance but not one which affects the picture of the basic physics.) Such alignment is not a sign of the breakdown of reflection invariance since spin (or any other form of angular momentum) is an axial vector which is permitted to be correlated with another axial vector (the magnetic field). However, what is happening in the decay of Co^{60} is that the emitted electrons are tending to align their momentum antiparallel to the spin of the nucleus. This is correlating an axial vector (spin) with a polar vector (momentum), which is not a reflection-invariant thing to do.

Reflection invariance provides an example of one of the puzzling features of particle physics. This is the selective satisfaction of symmetry principles. Look in a mirror and as far as strong or electromagnetic interactions are concerned, you will see the same laws as would be yielded by direct observation. Not so for weak interactions. Why some interactions should have a symmetry whilst others do not is something which we do not as yet understand.

One of the slogans of this chapter is that symmetry principles imply conservation laws. Therefore it is natural to ask what conservation law is valid for those interactions (strong and electromagnetic) which are reflection invariant. The answer is an interesting one. It turns on the fact that reflection invariance is what is called a discrete symmetry, as opposed to translational or rotational invariance, which are continuous symmetries. What is meant by that is as follows. If I go from A to B I can do so in steps, each as small as I please. If I make a rotation through an angle about some given direction again I can do so as gradually as I please. Thus these operations are characterized by the fact that they can be achieved in a Fabian fashion of small steps at a time. This smooth transition is what is meant by a continuous symmetry. Reflections are altogether different. There is a sudden change involved in going from direct observation to looking in the mirror, characterized by the fact that what is right-handed in one picture becomes left-handed in the other. There is no halfway house between right-hand and left-hand. The change from one to the other is a discontinuous act. This is what is meant by calling reflection a discrete symmetry.

It turns out that continuous symmetries lead to what are called additive conservation laws. The conservation of energy or momentum or angular momentum all involve such laws, because to get the total momentum (say) of a system one adds together the amounts of momentum present in the parts which make it up, and similarly for energy or angular momentum. Six units here and four units there, so we have ten units in all. Discrete symmetries, however, correspond to multiplicative conservation laws. Such laws deal in quantities which can only take two values, +1 or −1. The quantity of this type which is associated with reflections is called *parity*. (It is too difficult to give its technical definition since this requires some knowledge of the formalism of quantum mechanics. Once again I must ask the reader to trust what I say.) To find the parity of a system made up of several parts you *multiply* together the parities belonging to these separate parts. In this way you still get an answer which is either +1 or −1! This part has parity +1 and that part −1, so that together they give a system of parity $(+1)\times(-1)=-1$. Conserved quantities which are reckoned up in this multiplicative way are said, very reasonably, to correspond to multiplicative conservation laws. Interactions which are invariant under reflection are said to conserve parity.

τ–θ puzzle

Just as particles have an intrinsic angular momentum or spin, so they also have an intrinsic parity. It is another example of a quantum number. One can have positive-parity particles (like the proton or electron) with intrinsic parity +1 and negative-parity particles (like the pion) with intrinsic parity −1. If the decay of a particle is due to a parity-conserving interaction, its intrinsic parity can be determined by looking at the parity of the state composed of its decay products, since the two must be the same. The application of this idea to some weak decays (which we now know was a mistake since weak interactions do not conserve parity) led to such a confusing situation that at length Lee and Yang had the courage to suggest that maybe these interactions were not reflection invariant after all. The story went this way:

In the late 1940s and early 1950s people identified some new mesons which were almost four times heavier than the pion. These new particles were unstable with lifetimes (10^{-8} seconds to 10^{-10} seconds) which corresponded naturally to a weak decay process. Some of these particles, which were called θ-(theta) mesons, decayed into two pions, whilst others, called τ-(tau) mesons, decayed into three pions. Analyses of the decay products showed that the two-pion states were of positive parity and the three-pion states of negative parity. Since everyone assumed that parity was conserved, it was natural to think that θ and τ were different from each other and in particular had positive and negative intrinsic parity respectively. However, as other properties of these particles were investigated more closely, unease set in. This was because in all other respects the θs and the τs seemed so alike. In particular, as their masses got more and more accurately measured, so it seemed clearer and clearer that the values were the same. This could scarcely be just a coincidence and all sorts of ingenious theories were invented to try to explain why two different particles, with presumably two different sets of interactions, should happen to have the same mass. The ingenuity of these attempts exceeded their credibility. Eventually Lee and Yang made the brilliantly simple observation that perhaps the θ and the τ were the same particle, the two labels simply corresponding to two different ways in which it could decay. Since these two decay modes had opposite parity it was an inescapable consequence of this suggestion that parity is not conserved in the weak interactions responsible for the decays. Of course this explanation proved the correct one and the θ and τ were subsumed under the single name of K-meson. We shall return to the kaons in the next chapter.

C and T

The discrete operation of reflection is often denoted by P (for *parity*). There are two other discrete operations whose acquaintance we should make, which are denoted by C and T respectively.

C stands for *charge conjugation*. Under this operation we make the discontinuous change of replacing particles by antiparticles and vice versa. Thus under charge conjugation a hydrogen atom, with its positively-charged proton nucleus and encircling negatively-charged electron, turns into an anti-hydrogen atom, with a negatively-charged antiproton encircled by a positively-charged positron. Because C is a discrete symmetry there is a multiplicative quantum number, taking values ± 1 only, associated with it. It turns out that this quantity, called charge conjugation parity, is also not conserved by weak interactions, though it is conserved in strong and electromagnetic processes.

T denotes temporal reflection, or *time reversal*. This is the most difficult operation to understand, both for professionals and for amateurs.

Our whole experience of time is certainly unsymmetric. We can recall the past but we cannot foretell the future. The flow of time, therefore, appears to have a definite direction. The fact is sometimes referred to by the phrase 'the arrow of time'. Another way of expressing the effect would be to consider a film taken of some process and to ask whether we can tell the difference between the film being run forwards (ordinary time) or backwards (time reversed). For everyday processes we certainly can. A film of a man's life in which he gets noticably younger is being run backwards. A film of a bouncing ball in which the height of bounces gets smaller and smaller is being run forwards; if the bounces get bigger and bigger it is being run backwards. It is a surprising fact that these sorts of effects are not believed to indicate that the fundamental laws of physics have a built-in direction for time, but rather they are due to our looking at large and complicated systems. For such systems there is a purely statistical tendency for things to get more and more randomly mixed up as time goes on. This is the basis of the celebrated second law of thermodynamics which says that entropy (a measure of the randomness, or lack of structure of things) increases. The television advert in which the higgledy-piggledy pile of sweets jumps up and neatly rearranges itself in a careful array in the carton must be a film being run backwards because it shows this reverse trend from disorder to order that is not there in real life.

When we get down to the fundamental level of elementary particles the picture is different. As far as strong and electromagnetic interactions are concerned the laws are invariant under *T*. That is to say, we cannot tell the difference between a film of a strong or electromagnetic interaction event run backwards or forwards. That is not to say that in the laboratory a process may not occur more frequently in one direction than another. We often see a neutral pion decay into two photons,

$$\pi^0 \rightarrow \gamma + \gamma, \tag{A}$$

and seldom see two photons combine to give a pion,

$$\gamma + \gamma \rightarrow \pi^0. \tag{B}$$

The rate of occurrence depends on the number of different final configurations which are possible (technically this is called *phase space*). It is much easier to balance a walking stick on the end of one's nose and let it fall than to throw it up and catch it on one's nose. This is because there are many different ways in which it can fall but only one way in which it can land on the nose and stay balanced. Nevertheless one could not tell the difference between the balancing act and a film of the throw run backwards. Another example of the effects of phase space is provided by parallel parking, tricky to drive in (because one is trying to get to a unique position) and easy to drive out (because all sorts of final positions are possible), but certainly one motion is just the other reversed. In the same way it is phase space that makes event (A) common and (B) rare, but that does not prevent a film of (B) run backwards being indistinguishable from (A).

By now it will come as no surprise to learn that weak interactions are not quite invariant under the operation T.

It may be useful to pause for a moment to review our present state of knowledge. We have met three discrete operations:

P, parity (physics in a mirror);
C, charge conjugation (matter–antimatter interchange);
T, time reversal (the film run backwards).

Each operation leaves strong and electromagnetic interactions unchanged, so that for each operation there is a multiplicative quantum number, taking values +1 or −1 only, which is conserved by these interactions. None of the operations leaves the weak interactions unaltered, so that none of these quantities is conserved in weak decays. (However, it turns out that T is very nearly so.)

CP and TCP

While Lee and Yang had, as it were, shown that the physics of looking-glass land is weakly different from the physics of this side of the mantelpiece, people still retained a hankering after some sort of even-handedness in nature. For a while this seemed to be provided by the idea of CP conservation. CP is simply the operation of applying parity and charge conjugation simultaneously. If processes are unchanged by the combination of C and P it means that the same laws of physics are found if we observe a system of matter directly or an antimatter system reflected in a mirror. Somewhat oddly, it seemed for a while that weak interactions had this property. The thought was for some consoling, since it meant that in terms of fundamental physics there was still even-handedness provided you equated right-handed matter with left-handed antimatter.

Alas, the consolation was short lived! In 1964 the Americans Fitch and Cronin (in one of the most important discoveries still at the time of writing unrecognized by the Nobel Committee) found that there was a small effect, at about $\frac{1}{10}$% of the weak interaction strength, which violated CP. Its origin is still not properly understood.

It might appear from all this that at the level of weak interactions the state of discrete symmetries is pretty chaotic. So it is, except for one remarkable property. They are invariant under TCP, that is the simultaneous application of parity, charge conjugation, and time reversal. This means that, even for weak interactions, we cannot tell the difference between a matter system directly observed and a film of an antimatter system being run backwards and viewed in a mirror! Even more interesting is the status of this result. So far we have been talking about a range of possibilities. Interactions could be invariant under reflection or not. We could write down theories which had this property and others which did not. The choice of which to take could be settled only by experiment. Thus strong and electromagnetic interactions are found to be invariant under P whilst weak interactions are not. Similarly for C and T and for the pairs of products CP, CT, and PT. However, when we come to TCP everything is different. We do not know how to write down any theories which are not invariant under TCP. All the possibilities we can think of have this property. This result, which is called the TCP theorem, is, like the relation between spin-and-statistics of Chapter I, a general consequence of the combination of quantum mechanics and relativity (see Chapter V). Thus if the weak interactions were found to have some trace of TCP violation, something would have been discovered much more fundamental than the discovery of a small amount of CP violation. The latter, important though it was, simply meant we had to adjust the parameters of our theory a little to accommodate it. The former would mean that our basic principles were somewhere at fault. It would involve not some changes in the details of a theory, but a major conceptual overhaul of our approach to microscopic phenomena.

IV

The Plot Thickens

The symmetries we discussed in the preceding chapter are all connected with properties in space and time. They concern what happens when you move from point to point, rotate, reflect, consider earlier or later times, and so on. The only exception to this is charge conjugation, the exchange of matter and antimatter, but even that is related to space-time operations via the TCP theorem. Such symmetries are easy to picture and their character is clear. We understand them rather well. They correspond to statements about the properties of the world: no preferred special points in space or moments in time; no special directions; where appropriate, even-handedness; and so on. In terms of the dramatic metaphor to which we have recourse from time to time they are scene-setting statements about the stage on which the particle play is to be performed.

In this chapter we shall be concerned with relationships between the actors in the play, the particles themselves. We have already encountered the paradigmatic example of such a relationship in the symmetry between protons and neutrons as far as the strong interactions are concerned, which leads us to think of them as different modes of a single object, the nucleon. It is characteristic of the symmetries of this chapter that they are far more selective even than the discrete symmetries, P, C, T, of Chapter III. The latter were not respected by the weak interactions. Most of those symmetries with which we are now concerned refer only to strong interactions, and many of them only to some dominant part of the strong interactions. They are more difficult to picture than symmetries associated with space-time and their origin is correspondingly obscure. Nevertheless, the recognition and interpretation of such symmetries have provided many of the most interesting advances in the past 30 years in our understanding of the basic structure of matter. Before we can enter into this story we have to give an account of the experimental discoveries which were its starting point.

Strange particles

After 1945 people returned to pure science again. Accelerators were in their infancy and consequently many experiments had recourse to the bounty of nature in the form of the cosmic rays which bombard the earth from space outside it. The very energetic particles contained in cosmic rays mostly interact high up in the earth's atmosphere, so if their effects are to be studied it is necessary to send up detecting apparatus (a stack of layers of photographic emulsion in which charged particles would make tracks) to find out what is happening. Balloons provided the most popular means of transport. In this way a number of particles were first discovered, of which the θ and τ mesons of Chapter III are examples. Later these particles were artificially produced in the laboratory using accelerators. As their properties became better known a puzzle emerged. The new particles were quite copiously produced, which meant that there must be some powerful force responsible for their generation. In other words the new particles must be hadrons, having strong interactions with those hadrons already known, the nucleons and pions. On the other hand, by the standards of the microworld they lived a long time, so that their eventual decays were clearly to be attributed to weak interactions similar to those responsible for the decay of $\pi^{\pm}$ (see p. 21). It was not at all clear why this should be the case. For example, consider the θ decay mode of a K^+ meson into two pions,

$$K^+ \rightarrow \pi^+ + \pi^0.$$

All the particles here are hadrons and there is plenty of energy to make the process go since the K meson is nearly four times as heavy as a single pion. Why should the decay then not proceed as a strong interaction, giving a lifetime of about 10^{-23} seconds rather than the observed 10^{-8} seconds? Clearly something is inhibiting such a strong interaction taking place. The answer to the problem was given by the American theoretical physicist Murray Gell-Mann. The one way of slowing the process would be to find a conservation law respected by the strong interactions which the decay violated. No known conservation law had this property, so Gell-Mann boldly proposed that there must be a hitherto unknown quantity (or quantum number, as we say) whose conservation would do the trick. I regret to say that he called this new quantum number *strangeness*. This was the first step down the slippery slope of facetious terminology and fossilized jokes which have become endemic in the terminology of particle physics. We shall encounter many more examples before our story is ended. Great and creative as Murray Gell-Mann's influence has been on particle physics, in this one respect it has been a bane. The fact is the more surprising since Gell-Mann is deeply interested in linguistic questions as well as questions of physics.

Anyway, the deed was done and strangeness made its appearance. The proposal was that all the hitherto known particles (the particles of Chapter I) had zero strangeness. That explained why nobody had known about strangeness before; it hadn't been around. However, some of the new particles were to have non-zero strangeness. For example the K^+ would have strangeness $S = 1$. The conservation of strangeness in strong interactions then prevents its decay taking place as a strong process, since the pions both have $S = 0$.

Strangeness is another example of a quantum number, an intrinsic property associated with particles. In some ways it is like a new sort of charge, but distributed to particles in a way different from that in which electric charge is assigned, so that some electrically-charged particles have zero strangeness and some electrically-neutral particles have non-zero strangeness. However, there are very important differences between strangeness and charge quite apart from the way in which they are distributed. For one thing we believe that electric charge is absolutely conserved. No process has ever been seen not conserving it. Not so for strangeness. The decay of K^+ is not absolutely forbidden; it is only restricted to occurring as a weak interaction. Thus we encounter another example of selectivity in particle physics. The strong interactions, and also, as it turns out the electromagnetic interactions, conserve strangeness; the weak interactions do not.

Electric charge is more than just an accounting quantity, to be reckoned up and balanced out in interactions. It is also what we call a *coupling constant*, that is it measures the intensity of a particle's interaction with something, in this case with the photons of the electromagnetic field. This is, of course, the way in which particles' charges are actually determined. When we encounter gauge theories in Chapter VII we shall find a setting in which quantities like strangeness are related to coupling constants of strong interactions. Nevertheless, strangeness will not acquire the same physical directness that electric charge enjoys. Rather it is to be thought of as an attribute fulfilling the arithmetic role that Gell-Mann assigned to it, manifesting itself principally by forbidding this and allowing that.

In the laboratory in those early days one always began an experiment with ordinary or non-strange particles. For example, a proton collided with a proton. The conservation of strangeness in strong interactions then meant that strange particles would have to be produced in pairs, an $S = +1$ particle with an $S = -1$ particle so that their strangenesses cancelled out. This associated production of pairs of new particles was indeed found to be what was happening. Positive strangeness particles are the K^+ and K^0 mesons, negative strangeness particles include the antiparticles of the kaons, K^- and $\overline{K}^0$ (the latter different from K^0 because of its opposite sign of strangeness), together with some $S = -1$ baryons which are strange brothers of the nucleons. These latter consisted of a lone lambda particle, Λ^0, and three sigma particles, Σ^+, Σ^0, Σ^-. We say that the Λ is a singlet whilst the Σs form a triplet (just like the pions). The strange baryons are all heavier than the nucleon and decay into it with the emission of other particles; for example

$$\Lambda^0 \rightarrow p + \pi^-.$$

Since this does not conserve strangeness it must be a weak interaction and this gives the lambda a lifetime of 10^{-10} seconds. $\Sigma^\pm$ have weak decays also, but the Σ^0 has a strangeness-conserving electromagnetic decay into a lambda,

$$\Sigma^0 \rightarrow \Lambda^0 + \gamma,$$

with a lifetime of 10^{-14} seconds. All the decays illustrate another conservation law believed to be absolute. It is the conservation of *baryon number.* Baryons may change into each other but a state which starts with a baryon must also end with one, of one kind or another.

Resonances

So far all the particles we have met have been stable or (by our microscopic standards) long lived, having weak or electromagnetic decays. Lifetimes of 10^{-8} seconds, or even somewhat shorter, are long enough for charged particles to make tracks in bubble chambers or other detecting devices, so that the existence of these particles is visibly demonstrable. However, we should not exclude the existence of particles which can decay by strong interactions. They would have lifetimes of 10^{-23} seconds, which is far too short for any overt record of their existence to be available. Nevertheless many such particles are known. The first to be discovered were the Δs (delta). There are, in fact, four of them, with different electric charges, Δ^{++}, Δ^{+}, Δ^{0}, Δ^{-}, forming a four-fold multiplet in analogy to the three-fold multiplets of pions or Σs. The Δs have zero strangeness and so can suffer strong decays into nucleons and pions, like the decay of the doubly charged Δ^{++},

$$\Delta^{++} \rightarrow p + \pi^{+}.$$

The way they are detected is as follows. Suppose one is scattering a π^{+} on a proton. Then part of the time the π^{+} and the proton may like to stick together to form a Δ^{++}. This can only happen if the total energy of the π^{+} and p is equal to the mass of Δ^{++}. However, the latter is not quite an exact quantity. This is because there is an uncertainty relation (see Chapter V) between time and energy, or equivalently time and mass. Since the Δ^{++} only lives for a short time, its mass is consequently somewhat uncertain—it can only be said to lie within a range of values. (This is true for all unstable particles but if their lifetimes are long, the range of possible mass values is so narrow that to all intents and purposes there is a unique mass.) Thus in a certain limited energy range it is possible for the π^{+} and p to combine to form a Δ^{++}. After 10^{-23} seconds the Δ^{++} decays into a π^{+} and p again. What is going on is thus a two-step interaction of pion and proton,

$$\pi^{+} + p \rightarrow \Delta^{++} \rightarrow \pi^{+} + p,$$

proceeding through an unstable intermediate state, the Δ^{++}. Such processes are very familiar in nuclear physics where a projectile may stick to the target nucleus to produce an unstable intermediate state called the compound nucleus.

It turns out that in an energy region where such a two-step process is possible it is very much the dominant effect. It considerably enhances the *cross-section.* This quantity is a measure of how much scattering is going on. The bigger particles appear to each other, the more likely they are to hit each other and so to scatter. The cross-section is measured in terms of the area seen broadside on which one particle must present to the other to produce the amount of scatter it does. The larger the cross-section, the more scattering takes place. The chosen unit is the barn (10^{-24} cm^2) so called because on the microworld scale such an area does really present a target 'as big as a barn'. Typical particle cross-sections are not as big as that, ranging from a few hundredths of a barn downwards.

The enhancement of a cross-section due to an intermediate unstable particle is called a *resonance.* It is analogous to what one observes if one agitates one end of a taut string. If the end is moved up and down with a frequency corresponding to one of the natural frequencies of vibration of the string, then the induced effects are very large. In a similar way the mass ranges of objects like the Δ^{++} correspond to what one might think of as natural energies of the π^+-p system. When one is near them the consequences are spectacular.

Thus short-lived particles with strong decays, like the Δs, can be discovered by looking for enhancements in scattering processes, or bumps in the cross-section as we say. The position of the bump gives the average mass of the resonance, the width of the bump is a measure of its lifetime (the shorter the lifetime the wider the bump), the connection between width ΔE, and lifetime τ being given by the uncertainty relation

$$\tau = \hbar/\Delta E.$$

If that equation seems a rather mysterious hieroglyph, be patient. We shall discuss such matters in Chapter V.

There is no reason for treating unstable particles as any less particles than their stable or nearly stable brothers. The reason that the Δ^{++} is a resonance and the proton a stable particle is the accident of their different masses. Because the Δ^{++} is more massive than the sum of proton and pion masses, its strong decay is possible. Were it less massive it would be stable and if it had a mass less than the proton mass minus the pion mass then it would be the proton which would become unstable!

From a fundamental point of view resonances are particles just as much as anything else and in fact most of the particles now known to us are resonances.

Order out of chaos

During the 1950s and 1960s the number of 'elementary particles' increased enormously as more and more resonances were discovered. Soon the joke answer to the question of how many there were became 'ninety two', implying that the particle physicists were no better off than the chemists of a preceding generation with their 92 elements. Clearly such a plethora of particles was most unsatisfactory as a picture of the basic structure of matter. It was time to peel another layer off the onion. Such an operation is invariably a two-stage process; first some structure is discovered in the apparent chaos of 'elementary' objects, and secondly this structure provides a clue to the nature of the more truly elementary sub-particles of matter which constitute the next layer. Therefore, the first step was to try to find some sort of groupings of the known particles which would sort them out in an orderly fashion.

We have already taken the first step in that direction by recognizing the existence of sets of particles (nucleons, pions, Σs, Δs, etc) with similar properties under strong interactions but differing electric charges. This hadronic multiplet structure (to use the jargon) is a very important clue, and such groupings prove universal on the strongly interacting particle scene. The phenomenon is called *isotopic spin.* The reason for this name is not easy to explain but its importance makes an attempt worthwhile.

We are encountering a situation in which several objects have similar behaviour and it is natural to try to express this mathematically in terms of symmetries between them. This could be manifested by there being certain operations which interchanged the objects but which (because the objects are similar) made no real difference to the situation—'plus ça change, plus c'est la même chose'. For example, a circle is symmetrical under rotations about its centre because these just change round points on the circumference which are all equivalent to each other anyway.

Let us pursue the example of symmetry under rotations further. Suppose I choose conventional coordinate axes x, y, z, in three-dimensional space and suppose further that there are no preferred directions so that I have symmetry under rotations. Then the x and y and z axes are each as good as the other. In other words, if the system is invariant under rotations (no preferred directions) then the coordinates x, y, z are all on an equal footing and they therefore provide an example of a symmetric triplet of objects. The mathematicians express this by saying that the vector whose components are (x, y, z) is a *representation* of the three-dimensional rotation group. For the reader with a little (modern) mathematics I shall try to explain this idea in a bit more detail in an appendix at the end of the chapter. The non-mathematical reader will have to accept on trust the following set of statements:

(i) There are mathematical objects called Lie (fortunately pronounced 'Lee') groups (paradigmatic example: rotations) which are natural ways of expressing symmetries.

(ii) Associated with Lie groups are representations (paradigmatic example for the rotations in three dimensions: the three components x, y, z of a coordinate vector) which give sets of objects that are equivalent to each other under the symmetry of the Lie group and which therefore provide explicit examples of symmetrical multiplets. We shall want to associate these mathematical constructions with the concrete multiplets provided by sets of similar particles.

All this may seem a bit demanding. What it boils down to is this: mathematics provides an elegant machinery for constructing symmetrical sets of objects. Let us see if we can put this machinery to work in the service of physics.

One of the first questions to be determined in this programme is what is the appropriate Lie group for generating the symmetry we need for the physical word. When we are considering nucleons, pions, deltas, etc. the answer proves to be just our original friend, the rotation group in three dimensions. It is crucially important, however, to distinguish this use of rotations from that discussed in Chapter III. There we were thinking about the real-life three-dimensional space in which we and the particles move. Hence the rotations were honest rotations connected with facing now this way, now that. The three-dimensional space of isotopic spin has *nothing whatever to do with that.* It is a *fictitious* space that we introduce as a mathematical device for generating patterns of symmetry which will correspond to the multiplets of observed particles. The rotations in the fictitious space do things like change a proton into a neutron or a π^+ into a π^-. No amount of pirouetting in the real spatial world would produce that transformation. This point is so important that I must say it again. In trying to bring order into the elementary particle chaos we are going to put together particles with similar properties. Such symmetrical sets are naturally generated by mathematical systems called Lie groups. The particular version we are thinking of at the moment chooses the particular group which has the structure of rotations in some three-dimensional space. This three-dimensional space is *not* the one in which we live but is simply a mathematical artefact, useful because it generates symmetry sets of the right type for our present purpose. If you like, it is similar to using spatial metaphors to describe non-spatial situations. If we say that so-and-so is at the top of his profession we do not mean that he literally stands on the shoulders of his colleagues.

We can emphasize this fact by throwing away the fictitious three-dimensional space altogether. This is possible because the group of three-dimensional rotations is isomorphic to another group called SU(2). Isomorphism means that two mathematical objects which are generated in apparently different ways have in fact identical structure. In consequence they lead to exactly similar symmetry patterns; each is as good as the other for that task. Thus SU(2) is as suitable for multiplet purposes as the rotations we first thought of. (Technically SU(2) corresponds to the group of 2×2 unitary matrices of determinant one—the special (S) unimodular (U) group of order two (2).) Our reason for turning to SU(2) is not just to make sure we don't get confused about real and fictional spaces to rotate in. More important is that SU(2) contains the seeds of future development; after SU(2) comes SU(3) and beyond, and that is the way we shall have to tread.

Before pursuing that question it would be tidy to deal with a point of nomenclature. Why isotopic spin? The 'spin' is an echo of the idea of three-dimensional rotations. The 'isotopic' bit is frankly a disastrous misnomer which has become hallowed by use. Isotopic spin transformations turn a proton into a neutron or vice versa. In a nucleus such a change produces isobars (nuclei of nearly the same mass but different atomic number) rather than isotopes (nuclei of the same atomic number but different mass, obtained by adding or subtracting neutrons and leaving the photons unchanged). Therefore logically isotopic spin should be called isobaric spin. The late Leon Rosenfeld when he edited the journal *Nuclear Physics* conducted a brave campaign to effect this change. Papers with the words 'isotopic spin' were rejected until the offending phrase was replaced by 'isobaric spin'. Everyone agreed on the sense of the new nomenclature; everyone equally preferred to use the old familiar term. Conservatism triumphed and isotopic spin it is.

SU(3)

We have made great progress by introducing the groupings of elementary particles which correspond to the notion of isotopic spin. Its application to the nucleons produces the charge-independent forces that we discussed in Chapter I. However, this SU(2) theory can hardly be thought to provide a wholly satisfactory classification scheme. It has two defects. One is that there are still too many multiplets for one to feel that an economic account of particle theory has been given. The other is that there are significant aspects of particle properties which remain unincorporated in the scheme. The most important of these is Gell-Mann's new quantum number, strangeness. Both of these problems are solved by going to a bigger Lie group to generate the symmetry; from SU(2) to SU(3) in fact. However, to do so is something of an act of courage. It happened in the following way.

It is no disrespect to Heisenberg to say that it was somewhat natural to put the proton and neutron together into a nucleon doublet. After all, they have almost the same mass and look so similar, electromagnetism always excepted. All the SU(2) multiplets have this somewhat inevitable character. As their number increased people began to be bolder in their speculation, driven by a desperate need not to let the proliferation of particles get out of hand. There is a set of baryons which at least have in common the fact that they all have spin $\frac{1}{2}$. This set consists of the nucleon doublet, the $S = -1$ Λ-singlet and Σ-triplet, and a doublet of particles we have not met before, the $S = -2$ Ξ^- and Ξ^0 (xi). The last are always called the cascade particles, not just because some theoretical physicists are not completely at home in the odder parts of the Greek alphabet, but also because the name aptly describes their properties. In the weak decays which violate strangeness conservation, there is a limit to the rate at which strangeness disappears. In fact it is lost only a unit at a time. Thus a doubly-strange particle like an Ξ has first to decay into a singly-strange particle, like a Λ or a Σ, which then loses a second unit of strangeness when it in turn decays into a nucleon. Thus one gets a two-step decay chain like

$$\Xi^- \rightarrow \pi^- + \Lambda^0$$
$$\hookrightarrow \pi^- + p,$$

which is called a cascade process.

Now the nucleons, Λ, Σs, and Ξs are not all that alike. They differ considerably in mass, the Ξs being some 40% heavier than the nucleons, with the Λ and Σs in between. Yet it is tempting to try to combine these four SU(2) multiplets into a bigger (and so better) single grouping of eight particles which we might call a supermultiplet. The boldness of the step lies in the fact that certainly the strong interactions must treat these particles to some extent differently, if only because their masses are so unlike. Therefore members of a supermultiplet are only *approximately* the same. Nevertheless the step is one of great fruitfulness.

To generate supermultiplets you need a ‘super’ or enlarged Lie Group. The answer proves to be SU(3) (technically the group of 3×3 unitary matrices of determinant 1) instead of SU(2). With hindsight that seems a pretty straightforward thing to try. At the time it was more difficult. Partly this was because most theoretical physicists then were pretty hazy about Lie Groups and did not understand them properly (a condition for which the reader may have some sympathy). The other reason was simply the greatness of the step involved in linking together particles with masses as disparate as those of the nucleons and the cascades.

Important consequences flow from the transition to SU(3). The fact that it is a bigger group enables it to contain more information. Strangeness is now brought within the ambit of the Lie group approach. In fact it turns out that rather than strangeness itself a quantity called *hypercharge* (Y) compounded of strangeness and baryon number (B) in equal proportions,

$$Y = B + S,$$

is the appropriate quantity. (Mesons are conventionally assigned $B = 0$.) Since baryon number is always conserved, interactions which conserve S will also conserve Y, and those which allow S to change by one will allow Y to change likewise. The reason for preferring hypercharge is that it incorporates into the mathematics more neatly than strangeness (technically it is one of the operators of the Lie algebra). Furthermore, since SU(3) contains within itself all the structure associated with SU(2), it also specifies what isotopic spin properties are to be expected within the supermultiplet. All in all SU(3) is a much more predictive and much more constraining symmetry to work with than SU(2). The contrast is marked. For example, SU(2) has representations of any multiplicity. That means that as far as SU(2) is concerned any number of particles can be associated with one of its symmetry sets. One particle (Λ^0), two particles (nucleons), three particles (Σs), four particles (Δs); you name it, whatever number of particles you find looking the same as each other there is always an isotopic spin multiplet which will contain them. SU(2) is a very accommodating group, too accommodating in fact for physicists with a thirst for prediction. Not so SU(3). Only certain numbers of objects can be associated in one of its representations. For example, there is no nine-fold representation of SU(3). That implies that if you find nine particles looking like each other in the supermultiplet sense either you had better find another partner to make the number up to ten (for which there is an SU(3) representation) or SU(3) is no good to you after all. Of course such constraints are of the essence of a good physical theory. They provide predictive bite, leading either to triumphant vindication or summary dismissal.

Not only does SU(3) lay down certain possibilities for the number of particles associated together in a multiplet, but it also specifies their properties rather closely. This is because, as we have noted, it contains within itself hypercharge and isotopic spin and so it determines what values these quantities can take. There is a representation of SU(3) which contains eight objects. This

$Y=+1$	X X (doublet)		p n
$Y=0$	(triplet) X X X X (singlet)	$\Leftrightarrow$	Σ^+ Σ^0 Σ^- Λ^0
$Y=-1$	X X (doublet)		Ξ^0 Ξ^-
	Mathematics		Physics

octet cannot escape consisting of two isotopic doublets of hypercharges $Y=\pm 1$ respectively, together with a singlet and triplet of hypercharge zero. This is *exactly* what is needed to correspond to the nucleons ($Y=1$), Λ and Σs ($Y=0$), and cascades ($Y=-1$). SU(3) fits this baryon supermultiplet like a glove. Likewise it fits an analogous octet of mesons consisting of the K^+, K^0 doublet with $Y=1$, the antikaon doublet K^-, $\overline{K}^0$ with $Y=-1$, together with $Y=0$ members which are the pion triplet and a singlet, the η (eta), which is a heavy neutral meson whose acquaintance we have not made before. So remarkable are these successes that a euphoric Gell-Mann could be forgiven his punning reference to the 'eight-fold way'. These examples typify the great success of SU(3) as a pattern-making symmetry group. All known sets of particles can be accommodated within its close confines.

As another example of SU(3) in action let us consider, in a form simplified by hindsight, the argument that led Murray Gell-Mann to predict the existence of a new particle, the Ω^- (omega minus), and so helped him on his way to the Nobel prize. It is particularly simple to present the pattern in visual form. In

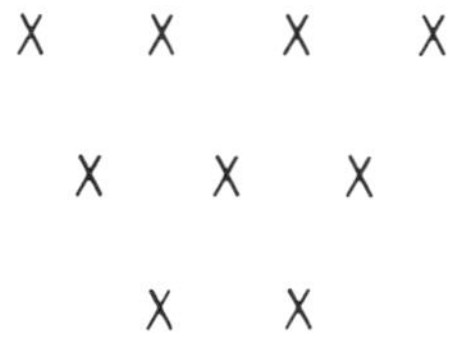

X X X X

X X X

X X

the figure the crosses represent particles arranged in rows corresponding to isotopic spin multiplets. All the particles are in fact short-lived resonances. The top row is the Δ-quartet, the second and third rows are triplets and doublets called Y* and Ξ^* respectively. The essential similarity that they have is that (in contrast with the spin-$\frac{1}{2}$ baryon octet we have been discussing) they all have spin $\frac{3}{2}$.

I do not know what age a child has to be to feel that there is something incomplete about this figure, but I do not suppose it needs to have reached very advanced years. Clearly there must be a final cross on a new line by itself to complete the triangular array. That is why there is a 10-fold representation of SU(3) but no 9-fold one. The prediction of that extra cross is the prediction of the existence of the Ω^-. However, we can go beyond that and say what some of its properties should be.

Each of the rows in the figure has different hypercharge. The first row has $Y=1$, the second $Y=0$, the third $Y=-1$. What then will be the hypercharge of the new row with the Ω^-? Of course the answer will be $Y=-2$, since we are obviously going down in steps of one. This means that the Ω^- is a baryon with strangeness -3 (if we remember that $Y=B+S$).

A slightly more subtle deduction relates to the mass of the Ω^-. The average mass of a Δ is 1232 in appropriate units (millions of electron-volts, actually). The average mass of a Y^* is 1385 and the average mass of a Ξ^* is 1530. To see what is happening a little subtraction is necessary:

$$1385-1232=153$$

$$1530-1385=145.$$

In other words the mass difference between neighbouring rows of the figure is almost the same value, 150 say. We can guess that this will hold for the final row also, which predicts a mass for the Ω^- of about

$$1530+150=1680.$$

In fact Ω^- when it was discovered was found to have a mass 1672, which is close enough.

The combination of these two predictions has a fascinating consequence which Gell-Mann recognized. The Ω^- is triply-strange ($S = -3$; very strange indeed, you might say). The simplest way in which it could break up in a strangeness-conserving strong interaction way would be to produce a doubly-strange Ξ and an antikaon,

$$\Omega^- \rightarrow \Xi^- + \overline{K}^0.$$

For this to be energetically possible the Ω^- would need to have a mass at least a little larger than the sum of the Ξ and $\overline{K}$ masses:

$$1321 + 497 = 1818.$$

With its actual mass of about 1680 such a strong decay is not possible. Thus, unlike the other members of its SU(3) decuplet, which are all resonances, the Ω^- is a long-lived particle with only weak decays. Because it is triply strange it is even more cascading than the Ξs, shedding its strangeness in unit steps, in sequences like

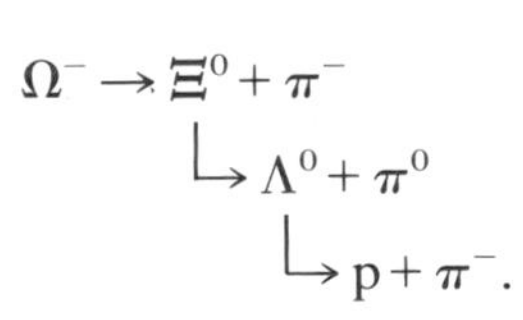

This striking behaviour helped in its recognition when the Ω^- was first found experimentally.

This discussion of the Ω^- may have made it seem only too easy to win a Nobel prize in particle physics. Remember that it is the depth of insight which can reduce important physical problems to such simple patterns as the figure from which we started which is being acknowledged in the award of such a prize.

The very success of SU(3) may dazzle us so much that we do not perceive what a profoundly curious thing is happening. Mathematics is a pure invention of the human mind, subject only to those canons of elegance and economy which mathematicians invoke when they describe something as beautiful. Lie groups are a prime example of such creativity. The pursuit of mathematics would be sufficiently justified by the austere but real pleasure that it affords in itself. Yet there is this astonishing extra reward, that some of its most elegant patterns are reproduced in the structure of the physical world. This most abstract of subjects is the one that holds the key to the universe. It is a truly remarkable fact. In Chapter X I shall try to say what I make of it.

Quarks

The advent of SU(3) produced a sufficient understanding of structure for it to be possible to attempt to penetrate to the next layer of the elementary particle onion. The breakthrough comes from taking the 3 of SU(3) seriously. In a hazy numerological way one could believe that this might indicate that some triad of objects would have a fundamental role to play. Those who have understood the aside about SU(3) being the group of 3×3 unimodular matrices (but don't worry if you are not one of them) will even more readily agree that the three-entry vectors on which these matrices act might be expected to have special status. Let us put it this way. SU(3) is concerned with shuffling operations performed on three objects. Is it not natural to suppose that these three objects have some basic role to play? These intimations are correct. Mathematically there is a triplet representation of SU(3) which is called the fundamental representation because all other representations, such as the octets and decuplets of the last section, can be constructed by mathematical manipulations on these three fundamental objects. The mathematical operations involved are direct products, symmetrizing, and making traceless. Let him who can understand take note. As for those who cannot, I apologize for this brief but mysterious extract from the mathematical recipe book. The essential point is that *mathematically* everything we have been talking about can be constructed out of three basic elements. There is no logical need for this mathematical fact to be paralleled by a corresponding physical statement that the particles are themselves constructed out of three basic constituents. We are treating Lie Groups as symmetry pattern-generators but there is no necessity to adopt so literal an interpretation of their role. (After all, in the classical theory of magnetism it is sometimes convenient to talk about magnetic poles, in analogy to electric charges, without thereby implying any expectation that isolated magnetic poles exist as physical objects.) Nevertheless, it was clearly an attractive notion that there might be this physical parallelism, for then we should indeed have reached a yet more fundamental level in the structure of matter. It was clearly an option that people would wish to explore.

A name was coined for these putative constituents by that great inventor of both physics and terminology, Murray Gell-Mann. This was too serious a subject for jokes. Instead he had recourse to an enigmatic line of James Joyce's *Finnegans Wake*:

'Three quarks for Muster Mark'.

These basic building blocks of matter are therefore called quarks. (The canonical pronunciation makes them rhyme with 'corks', though the deviant pronunciation rhyming with 'sparks' is also encountered.)

SU(3) fixes a number of properties of the quarks. Just as the octet and decuplet had components of specified isotopic spin and hypercharge properties, so also the triplet decomposes in a determined way. It turns out that the quarks must form an isotopic doublet of hypercharge $Y=\frac{1}{3}$ and an isotopic singlet of hypercharge $Y=-\frac{2}{3}$. The appearance of these fractions signals the most celebrated property of the quarks. There is a relation which holds between hypercharge, isotopic spin, and electric charge. It states that the maximum electric charge Q in an isotopic spin multiplet with $(2I+1)$ members and with hypercharge Y, is given by

$$Q=I+\tfrac{1}{2}Y,$$

the other charges in the multiplet lying in unit steps below this maximum value. It is easy to check that this gives the correct charges for the octets and decuplets. For example, the pair of nucleons have $I=\frac{1}{2}$ and $Y=1$, so that their possible charges are $Q=+1, 0$, as is indeed the case. Applying this to the quarks gave charges $\frac{2}{3}$ and $-\frac{1}{3}$ for the two members of the doublet (usually called u and d, or up and down quarks) and $-\frac{1}{3}$ again for the singlet (usually called s, or strange quark, or more facetiously, sideways quark). These were very odd values, for no one had ever seen a particle which did not have integral charge. That is, all the known particles had whole number charges like 1, 0, -1, and so on. Moreover, since charge is absolutely conserved and never gets lost anywhere, one of these fractionally charged quarks, if they exist, must be stable, for it is just impossible for it to turn into ordinary particles with integral charges. That means that quarks might be lying around anywhere. A feverish search began to find them, inspired no doubt partly by the thought that the first man who indubitably discovered one might find that he had also earned for himself an enjoyable trip to Stockholm. The technique for the search was essentially based on highly ingenious refinements of the idea used by Millikan in 1910 to measure the charge on the electron. The difference, of course, was that this time it was hoped that fractional answers would be found occasionally, corresponding to the presence of quarks. It was thought that cosmic rays would have been showering the earth with quarks for many aeons and people tried to work out where they might tend to accumulate in quark-rich deposits. The bottom of the ocean was a good bet and quantities of mud were sucked up to be submitted to analysis. Others reckoned that the earth's magnetic field would help to focus this rain of quarks into the polar regions, so that chunks of arctic ice would be worth investigating. Fractionally charged objects would have odd chemical properties and so another line of enquiry concentrated on how this might influence where to look. I seem to recall having read of one investigator who propounded the theory that quarks would accumulate in the muscles of bivalves and that he was provided with a mound of oysters with which to pursue this point. It is sad to relate that all this activity came to nothing. So far no one has found a bona fide quark.

Too bad! So another good idea comes to nothing? Particle physicists have been most reluctant to reach this conclusion. The first thing that gives them pause is the astonishing success of the quark model for the structure of matter. To get angular momentum properties right the quarks must have spin $\frac{1}{2}$. Then the mesons are most simply thought of as $q\bar{q}$ systems, that is the combination of a quark with an antiquark, which will have integral spin. Such a combination can be shown to give mesons which are members of a SU(3) octet or which are SU(3) singlets, the latter possibility corresponding to a single meson of isotopic spin zero and hypercharge $Y=0$. Just such a picture fits the known mesons perfectly. Baryons are then thought of as qqq systems, that is they are constructed out of three quarks, which gives half-odd-integral spin as required. This gives rise to baryon supermultiplets which are either singlets, octets, or decuplets.

Again that is exactly what is required to describe nature. Thus the quark model gives a pattern of strongly interacting particles, or a hadron spectroscopy as we say, which is just right. The $q\bar{q}$ and qqq combinations select precisely those representations of SU(3) which are the ones which work, no more, no less. (For example, the model does not permit mesons to be decuplets, nor are they found to be so.) Moreover it provides a simple picture of why SU(3) is not an exact symmetry. By giving the s-quark a different mass from the u and d quarks, we can reproduce the sort of mass splittings between particles of different hypercharges which are encountered experimentally.

It is hard to believe that so persuasive and successful a picture is not describing a physical reality. That feeling will be intensified when we describe deep inelastic scattering experiments in Chapter VII. It will then be time to face the dilemma that no one has seen a quark despite the evidence that everything is made of them. The answer proposed may indicate that we have reached the centre of the onion.

Subplot

Too many particles

↓

Multiplets = isotopic spin = SU(2)

\+

strangeness

↘

SU(3)

octets and decuplets,
'shuffles' *three quarks* (u, d, s)

Moral: There is more to mathematics than just pure thought. Its patterns are realized in the physical world.

Mathematical appendix

Lie groups are too interesting to leave without some attempt to convey a little more to the reader who has an elementary knowledge of mathematics. Those without such knowledge can hasten on to Chapter V.

In general terms a group is a mathematical system which contains the possibility of multiplication together with the existence of an identity element, that is something which (like 1 in ordinary numbers) does not change the quantity into which it is multiplied. We also require the existence of an inverse, that is for each element a there is another element a^{-1} such that $aa^{-1} = a^{-1}a = e$, where e is the identity. (We do not emphasize this latter point in what follows, but it is easy to check that our examples satisfy this further condition.) Such mathematical structures are much more widespread than the simple example provided by numbers with which we are all familiar. Rotations provide an example of a non-numerical group. We define multiplication of two rotations by performing one rotation followed by performing the other. (For rotations in a two-dimensional plane the order in which we choose to do the rotations does not matter; we end up in the same configuration either way round. The multiplication is in this case called commutative and the group is said to be abelian. For rotations in three dimensions the order does matter; in general we end up in a different configuration if we reverse the order. The multiplication in this case is said to be non-commutative and the group is non-abelian.) The identity is provided by the trivial rotation when we rotate through zero angle. Obviously this changes nothing. It is worth thinking about rotations in a two-dimensional plane in a bit more detail. In the figure we see that rotation through an angle θ changes the axes Ox, y

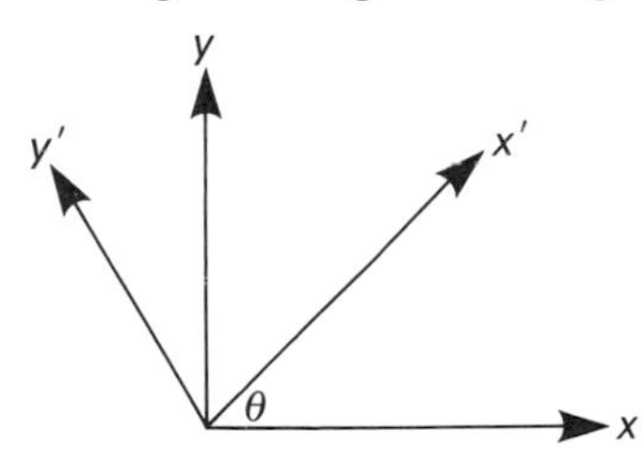

into the axes Ox', y' A little elementary trigonometry then shows that the relationships between these two sets of coordinates is

$$\begin{aligned} x &= x' \cos\theta - y' \sin\theta, \\ y &= x' \sin\theta + y' \cos\theta, \end{aligned} \tag{1}$$

which we can write in the matrix form

$$\begin{pmatrix} x \\ y \end{pmatrix} = \begin{pmatrix} \cos\theta & -\sin\theta \\ \sin\theta & \cos\theta \end{pmatrix} \begin{pmatrix} x' \\ y' \end{pmatrix}. \tag{2}$$

If we call the 2×2 matrix in (2) R_θ then it is easy to verify that multiplying two such matrices corresponding respectively to rotations though θ and ϕ gives the matrix corresponding to a rotation through $\theta+\phi$,

$$R_\theta R_\phi = R_\phi R_\theta = R_{\theta+\phi}. \tag{3}$$

Thus the two-component vector (x, y) generates a set of matrices R which multiply together *in exactly the same way that rotations multiply together.* This is why we call the vector (x, y) a representation of the two-dimensional rotation group; it gives matrices whose properties precisely mirror those of the rotations.

If we were to consider three-dimensional rotations the vector (x, y, z) would give a set of 3×3 matrices which represented rotations in three dimensions. The principle is similar, though the algebra and trigonometry are a bit more complicated. Contained within the group of rotations in three dimensions is a subgroup corresponding to two-dimensional rotations in the (x, y) plane only. The vector (x, y, z) splits into two when we consider this restricted subgroup of rotations. The (x, y) components will just correspond to the representation of two-dimensional rotations given by (2) and (3). The z component, on the other hand, is unaffected by rotations in the (x, y) plane. We say that it is invariant and it corresponds to a one-dimensional representation with $R=1$ for all θ, which trivially satisfies (3). This shows us an example of how a representation of a group (in this case the three-vector (x, y, z) representation of rotations in three dimensions) decomposes into a specified set of representations (in this case the two-vector (x, y) and the invariant z) of a subgroup (the two-dimensional rotations in the (x, y) plane). It is a similar sort of reduction which leads to an SU(3) representation containing within it a specified set of SU(2) representations (isotopic spin multiplets).

V

Dramaturgy

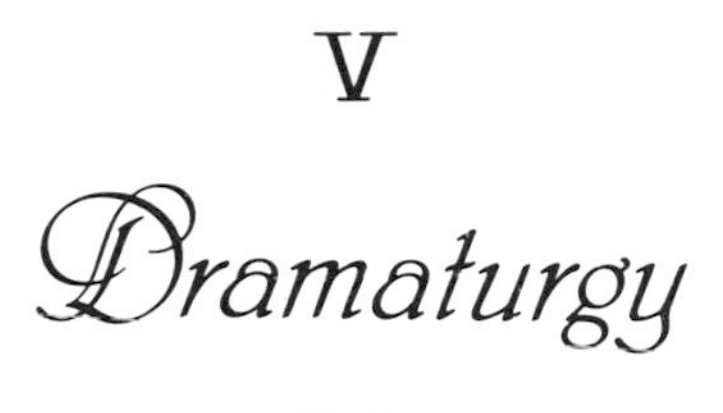

Our account of particle physics has so far concentrated on describing the different types of particles and the structural relations exhibited by arranging them into multiplets. Such an activity can be thought of in terms of our theatrical metaphor as corresponding to the compilation of cast lists together with some rudimentary elements of plot. However, we can go on to enquire what are the rules which will control the action on the stage, what are the counterparts for particle physics of the Aristotelian unities of space and time, or some similar prescription of dramaturgy? In physics terms, what are the dynamical laws of particle interactions? After all, what we have been engaged in up to now, despite its fascination, has been perilously close to natural history rather than science, a patient observation of remarkable regularities in nature. That, for sure, is where we need to start, but we must hope to finish with the full understanding which comes from the possession of a detailed theory. What form should such a theory take and what are the basic principles on which it should be founded?

The answer lies in the application of the two great discoveries which mark off twentieth century physics from what has gone before. Because elementary particles are small they require quantum mechanics to describe them. Because they move fast they are subject to special relativity. The fusion of these two principles to form relativistic quantum mechanics gives a theoretical basis which we believe is adequate for the phenomena of particle physics without the need of further new principles. Special relativity sprang fully armed from the brow of Einstein in 1905. Quantum mechanics was more gradually unravelled, largely from the investigations centring on the photon which we described in Chapter I. The fifty years which have followed the first uniting of these two principles have served to reveal what a rich and intricate theory was formed by their coming together. In this chapter we try to understand this in a little more detail. A proper treatment of the theme would obviously make technical demands far beyond the modest scope of this book, but nevertheless it is possible to provide something of the flavour of things without too great an oversimplification.

Quantum mechanics

We have already encountered one of the most characteristic effects of quantum theory: its replacement of the hitherto continuous by the discrete (see p. 6). Angular momentum, which classically can take any value, we found to be limited to multiples of $\frac{1}{2}\hbar$. A similar effect associated with simple harmonic oscillators proves the key to important developments.

A simple harmonic oscillator is a system which oscillates with a single frequency. A simple physical realization is provided by a pendulum. The oscillator provides a basic dynamical unit out of which more complicated systems can be built. Consider, for example, the electromagnetic field. Some part of electromagnetic radiation manifests itself as what we see as light. When white light is passed through a prism and split up into its component colours, it is being decomposed into different frequencies. The faster oscillations we see as the blue end of the spectrum, the slower as the red end. Another part of electromagnetic radiation forms radio waves. When we twiddle the tuning knob on our set we are sampling different frequencies of radiation, all oscillating very much less rapidly than those associated with visible light. In this way we can see that a complicated system like electromagnetic radiation can be decomposed into a set of frequencies and so be thought of as being equivalent to a vast collection of oscillators, one corresponding to each of the different frequencies present. Technically such a decomposition is called Fourier analysis. It has the consequence that if we understand the result of applying quantum mechanics to a single simple harmonic oscillator we shall also be able to understand its consequences for much more elaborate systems like the electromagnetic field.

Think of a pendulum. In the (classical) everyday world it seems that by drawing the bob aside an appropriate distance from its equilibrium position we can give the pendulum any energy we please. If the displacement is small, the energy is correspondingly tiny; if the displacement is large, the energy is large also. Continuous adjustment between these extremes appears possible. However, this classical continuity is in reality an illusion; quantum mechanically it is replaced by a discrete set of possible energies. An oscillator of frequency ω (that is, executing $\omega/2\pi$ oscillations per second) can only have an energy which is one of the set of values

$$E_n = (n + \tfrac{1}{2})\hbar\omega, \qquad n = 0, 1, 2, \ldots,$$

where $\hbar$ is Planck's constant and n a whole number. The reason that we don't notice this with our grandfather clock is, once again, because $\hbar$ is such a small unit. For the pendulum of a long-case clock n will be in excess of 10^{30}. It is another example of a row of closely-spaced dots looking like a continuous line.

The $\frac{1}{2}$ appearing in the formula for E_n looks rather odd. One might think it was a mistake. In fact, it corresponds to an important physical effect. It implies that the lowest energy that the oscillator can have is $\frac{1}{2}\hbar\omega$. Classically, of course, the lowest energy is zero, corresponding to the bob of the pendulum being at rest at its lowest point. However, quantum mechanically this state is not possible; the Heisenberg uncertainty principle forbids it.

The uncertainty principle encapsulates the essence of quantum mechanics for it lays down the limits of knowledge which are characteristic of the theory. Classical mechanics is played out before an all-seeing eye, for it supposes that all the time we know where particles are (their positions) and what they are doing (their momenta). Observing these quantities may disturb the system a little but Newtonian mechanics supposes it possible to reduce this disturbance to as small an amount as one pleases, so that it represents at worst a practical problem but never a difficulty of principle. Not so for quantum mechanics. Planck's constant $\hbar$ provides a scale which specifies an irreducible minimum of disturbance by the observer which cannot be avoided. If we measure a particle's position to an accuracy of Δx, then its momentum will be uncertain by at least an amount Δp, where

$$\Delta x \cdot \Delta p = \hbar.$$

Thus if we know where a particle is (Δx very small) we do not know what it is doing (Δp consequently large). Equally, if we know what it is doing (Δp very small) we do not know where it is (Δx consequently large). A similar uncertainty relation connects energy and time,

$$\Delta E \cdot \Delta t = \hbar,$$

which we referred to in our discussion of resonances in Chapter IV.

The reason that we can forget about Heisenberg in the everday world is, of course, the smallness of $\hbar$. Suppose we measure the position of a billiard ball to an accuracy of a millionth of a centimetre. Then its velocity is uncertain by an amount 10^{-23} cm per second! That is a degree of imprecision that we can certainly afford to live with. The case is very different for an electron. An uncertainty in its position of 10^{-10} cm requires an uncertainty in its velocity which amounts to a third of the velocity of light. That is a far from negligible effect.

People with philosophical inclinations and rather hazy notions of science sometimes tend to get over-excited about the uncertainty principle. No one would wish to deny its importance for small-scale physics but there is nothing mysterious about it. The limitations on accuracy which it implies arise from the existence of the photon. One way of seeing where something is is to shine a light on it. In bouncing off the object the light disturbs its state of motion somewhat. Because there is a smallest unit of light (a single photon), there is a corresponding minimal degree of disturbance. This will only matter for systems which are of a magnitude comparable with a photon; important therefore for electrons, but negligible for billiard balls.

The implications of all this for the simple harmonic oscillator are clear. In the classical zero-energy state we know where the bob is (at the bottom) and what it is doing (it is at rest). Heisenberg does not allow this for a quantum mechanical oscillator. Its lowest energy state must involve a judicious compromise with the requirements of the uncertainty principle; the bob almost at rest and near the bottom. The energy associated with the resulting slightly quivering system is the $\frac{1}{2}\hbar\omega$ given by E_n with $n = 0$. This quantum mechanical effect is called the *zero-point motion.* It is completely negligible for everyday pendulums, but in the microworld these effects are important.

The equation for E_n tells us that above the $\frac{1}{2}\hbar\omega$ of the zero-point motion the energy of the oscillator is reckoned up in terms of packets of energy of amount $\hbar\omega$. There may be none of these packets ($n = 0$), one packet ($n = 1$), two packets ($n = 2$), and so on. This immediately makes one think of photons. The presence of quantities reckoned up in integral multiples is the hallmark of a particle structure. The number n will be the number of such particles.

We have been describing the ingredients needed for the construction of what is technically known as *quantum field theory*. It provides a formalism which can accommodate the particle–wave duality of Chapter 1 without taint of paradox. First it is a *field theory*, that is it describes something (like the electromagnetic field) which varies from point to point and from time to time. This is what we need to describe wave motion, so that such a theory will have all the properties of interference which are characteristic of waves. Using Fourier analysis such a field theory can be expressed in terms of a set of oscillators. Because it is also a *quantum theory* it will then also contain within itself the discreteness, corresponding to the integers n in the formula for E_n, which is characteristic of particles.

On this interpretation the particles are the excitations (the extra amounts of energy corresponding to $n \neq 0$) which are present in the field. When there are no particles present (all the oscillators in their lowest states with $n = 0$) we shall have the vacuum. The zero-point motion means that this, contrary to our classical ideas, will not be a state with nothing happening. Rather, each oscillator will have its zero-point motion and zero-point energy $\frac{1}{2}\hbar\omega$. The field will contain many such oscillators (in fact an infinite number) so that this energy will add up to a large amount (in fact an infinite amount). This need not disturb us too much. In physical processes it is the energy *differences* which count. If I write down the conservation of energy for a process,

$$\text{Energy of initial state} = \text{Energy of final state},$$

I can subtract any number I like from both sides of this equation (even an infinite number if necessary!) and it still remains a true equation. This is what we do with the energy of the vacuum. We subtract it off from both sides so that in an energy-balance equation we need only reckon up the energies of the actual particles present. In this way it is possible to give the energy of the vacuum the satisfactory value zero.

While this is a legitimate procedure, it is very important to realize that subtracting off the zero-point energy does not mean that the phenomenon of zero-point motion is illusory. This cannot be so since we have seen that it is an inescapable consequence of the uncertainty principle. All we have done is a convenient book-keeping transaction with energy. It is rather like subtracting off allowable expenses from one's receipts before making a return of taxable income. The purposes for which those expenses were spent still remain as real activities.

Even in its lowest energy state a pendulum is quivering slightly. Applied many times over to the electromagnetic field this means that even when there are no photons present something is happening. An empty stage in particle physics is not one devoid of activity. This surprising state of affairs is called *vacuum fluctuations.* It manifests itself by the presence of transient fields, or fluctuations, in 'empty' space. On average, of course, these effects cancel out. However, if one takes a sample over a small volume for a short time then a significant fluctuation may make itself felt. The smaller the volume and the shorter the time, the more violent the effect is liable to be. This is particularly important for our point-like particles which are consequently all the time moving through a turbulent environment. Buffeted by vacuum fluctuations the elementary particle has a pretty rocky ride. This is the origin of the infinities which people found when they first tried to calculate with quantum field theories.

Before tackling that problem an attempt at consolidation might be worthwhile. Quantum field theory is pretty heady stuff, particularly at first acquaintance.

A field is something which varies from place to place and time to time. This automatically gives it wavelike properties. By splitting the field into its component frequencies (Fourier analysis) it can be treated as a collection of oscillators. Quantum mechanics applied to these oscillators introduces a discreteness (energy reckoned up in packets of $\hbar\omega$) which leads to a particle interpretation for the excitations of the field. Each field has its own particles. The electro magnetic field produces photons; other fields will be needed for quarks, electrons, muons, etc. Each field can be thought of as the potentiality for producing its particular type of particle, which appears as an actuality if energy is fed into the field to produce the necessary excitation. The phenomenon of zero-point motion means that even when no particles are present (the vacuum) the field is not completely quiescent. Vacuum fluctuations are inescapably there; pure potentiality is not devoid of activity.

Renormalization

The theory which initially attracted most attention was quantum electrodynamics, that is the theory which describes the interaction of electrons (or muons) with the electromagnetic field. The first problem was that the equations of the theory were too difficult to solve exactly. It was necessary to have recourse to some approximate treatment. The standard technique to use is called perturbation theory. You divide the problem up into two parts: one large, which you can calculate exactly; the other small in which the complications reside. Because of its smallness this second part is treated as a perturbation on the first. Suppose we were trying to determine the distance from A to B by a fairly straight road with a few bends in it. The simple soluble problem would be to treat the road as a straight line and measure the distance on the map. A more accurate calculation would have to take into account the more important wiggles as perturbations on that simple answer.

In applying this technique to quantum electrodynamics the simple part corresponds to describing the electrons and photons by themselves, without any interaction; the complicated part is the interaction between them, the emission or absorption of a photon by an electron. Of course the complicated part is also the interesting part. It is the interaction which we wish to study. We can treat it as a perturbation because it really is small. The strength of the interaction is characterized by a numerical parameter α, called for historical reasons the fine structure constant, whose value is tiny, approximately $\frac{1}{137}$. (The appearance of a whole number like 137 in the denominator is a numerical coincidence which beguiled the late Sir Arthur Eddington into curious speculative byways in a will-o'-the-wisp search for an explanation. In fact better measurements show that the denominator is not exactly a whole number, and it is best just to regard α as an experimentally determined constant of nature.) It is possible to calculate the effects of the interaction as a mathematical series in the fine structure constant. The first term is proportional to α; there is a further correction proportional to α^2 and so on. Since α is so tiny a number one might hope that this second (α^2) term would only represent a small addition compared to the first (α). Independently, in the second half of the 1940s, three people, the two Americans Feynman and Schwinger, and the Japanese Tomonaga, did calculations of this sort. However, there was a problem they first had to solve before they could get believable answers. When they worked out the numbers multiplying the powers of α in perturbation theory these coefficients turned out to be infinite! Some sense had to be made of this nonsensical situation.

As they wrestled with this problem, Feynman, Schwinger, and Tomonaga were each able to see that the infinities arose from the effect of vacuum fluctuations and that they were concentrated in two physical consequences. One related to the mass of the electron. From its interaction with the fluctuating electromagnetic fields in the vacuum, the electron acquires energy which contributed to its mass. If the electron is treated as a point particle this contribution is infinite. Secondly, and more subtly, the presence of fluctuating electron–positron pairs in the vacuum makes the vacuum behave like a polarizable medium and this modifies the apparent charge on the electron in the same way that charges give modified effects when placed in a dielectric medium. All infinities in quantum electrodynamics can be allocated to these two effects. This observation forms the basis of a bold procedure called *renormalization.* We have this ugly term including infinities which represents the electron's actual mass. All right, let us just replace it with the finite sensible value given by the measured mass of the electron. Equally we have this second ugly term including infinities which represents the electron's actual charge. Again let us replace it by the perfectly finite measured value of that charge. By these two strokes we have removed all infinities from quantum electrodynamics and left ourselves with perfectly well-defined predictions. The proof of the pudding in this sort of game is in the eating. Does it work? The answer is a triumphant 'yes'! Renormalized quantum electrodynamics is the one splendidly successful theory of the microworld. It predicts various effects (with names like the anomalous magnetic moment of the electron and the Lamb shift in hydrogen) which agree with astonishing accuracy with their measured values (that is to say, within the few parts per million which represent both the limits of experimental accuracy and of present calculations). For example, a quantity called the anomalous gyromagnetic ratio $(g-2)$ of the μ-meson is predicted to be

$$0.001165897 \pm 0.000000009$$

and found to be

$$0.001166160 \pm 0.000000310.$$

Theories in which the infinities can all be isolated in contributions to quantities like masses and charges are said to be renormalizable. These are the only theories with which we can begin to calculate sensible results. This inevitably gives them a special status. It is a matter of debate exactly how special this status is. The answer turns on one's assessment of the renormalization procedure.

Two views are possible. One simply says, *tout court*, renormalization is the only way of making sense of relativistic quantum mechanics, in which case renormalizable theories are the only theories we have got. The other view is more cautious. The infinities really arise from treating the electron, say, very literally as a point particle. If it were spread out a little or modified in some such way, the infinities would disappear. Now we know experimentally that electrons look point-like down to very small distances, of the order of 10^{-15} cm. If they are spread out at all it is over dimensions smaller than this. But really we have no clue about how the microworld behaves at distances which are very much shorter than 10^{-15} cm. Eventually gravitational effects begin to bend space-time and all sorts of queer things could happen. Perhaps these unknown effects actually make everything finite. Then the status of a renormalizable theory is rather different from the first austere view that it is the only possible type of theory. It is simply a theory in which the consequences of these unknown effects at very short distances are concentrated in just a few simple parameters, the masses and charges. We can adequately compensate for our ignorance of ultra small-scale physics by putting in the experimental values of these quantities.

A non-renormalizable theory would be one which could not be treated in this way and which therefore could only be made sense of when we really understood short-distance behaviour. Non-renormalizable theories would be abhorrent to deal with but not forbidden in principle.

While many great men appear to take the first point of view, for what it is worth I incline to the second. One consideration is that our knowledge of how to effect renormalization is tied closely to the sort of perturbation theory expansions of which the quantum electrodynamical series proportional to powers of α is the prime example. People are just beginning to learn how to think about field theories in ways that are different from this. It may be that in due course this will enlarge our range of sensible theories. Another consideration is that if quantum mechanics is applied to the field theory which is believed to describe gravitation (Einstein's theory of general relativity) then that theory does not appear to be renormalizable. We have explained that gravitational effects are too small to need to be taken into account in ordinary particle physics, but the interaction is nevertheless a fundamental one and we must surely believe that it is capable of consistent quantum mechanical treatment. Against these considerations must be set the fact, of which Chapters VII and VIII give an account, that renormalizable field theories, particularly in a form called gauge field theories, seem to provide just the tools we need for understanding particle physics.

Perturbation theory scored a great quantitative triumph with quantum electrodynamics because the fine structure constant α is so small a number and so two or three terms in the series will give very accurate answers. There is a second, qualitative, role which perturbation theory has played even in the understanding of hadronic dynamics, where the strength of the coupling precludes a literal interpretation of the series. (The parameter in strong interactions corresponding to α is 1 rather than $\frac{1}{137}$, so that the successive terms of the series do not get smaller.) This role is to provide a 'theoretical laboratory' in which ideas and conjectures about relativistic quantum mechanics can be tested. The successive terms which occur in perturbation theory can be written in a very neat mathematical form invented by Richard Feynman and called Feynman integrals. If each of these terms is found to possess a particular property then it is heuristically compelling (if not strictly mathematically inevitable) to suppose that this is a property of the theory itself. The intensive study of Feynman integrals has provided a valuable source of intuition, and a test-bed for conjecture, in relativistic quantum mechanics.

Relativity

Using the discrete properties of the energy levels of oscillators to introduce a particle structure into a theory is a trick which can be applied to non-relativistic or to relativistic systems. In fact field theory techniques can be applied to such systems as crystals to describe excitations called phonons (since they are 'particles' of sound in the way that photons are particles of light). However, as we have already emphasized several times, it is the marriage of field theory and special relativity which proves particularly fertile. Fruits of the union include the general principles of the relation of spin-and-statistics (Chapter I) and the TCP theorem (Chapter III). Unfortunately the proofs of these results do not lend themselves to the broad-brush treatment of popular exposition and their validity must remain a matter for trust between author and reader, rather than demonstration.

High energy elementary particles are moving fast, that is with velocities which are appreciable fractions of the velocity of light. They therefore provide excellent opportunities to exemplify characteristic effects associated with relativity. We shall be content to illustrate this by considering the celebrated prediction that moving clocks run slow compared with clocks which are at rest. It provides the explanation of the fact that high energy μ-mesons, created in the upper atmosphere by the decay of fast pions (which are themselves the products of primary cosmic rays) can travel several kilometres though the atmosphere to be detected by ground stations. The muon has a lifetime of 2.2×10^{-6} seconds, which would appear to mean that even if it were moving at a speed near the velocity of light the μ-meson could only travel a few hundred metres before decaying. However, the clock which tells the muon when to decay is, as it were, its own personal clock travelling with it. Since the muon is moving fast its clock will tick slowly compared with the clock of a stationary observer on the earth's surface. Thus to such an observer the muon's lifetime will appear greatly extended, giving it time to travel over distances of kilometres.

Four dimensions
(not for the mathematically squeamish)

Because of this tardy property of moving clocks special relativity abolished the old persuasive Newtonian notion of a uniformly flowing universal time. It was in this fact, and the consequent recognition that the simultaneity of two events did not have an absolute meaning but depended on who was the judge of it, that lay the daring of Einstein's genius. These ideas make it natural to think of space and time not as two separate entities but combined to form a four-dimensional space-time. Nothing mysterious is involved here. In particular, the distinctive characters of space and time are preserved. It would be absurd to deny that history is different from geography! The way it all works is that events have four coordinates assigned to them, the three space coordinates (x, y, z) in some spatial coordinate system, and a time t measured by a clock at rest in this coordinate system. What is significant is the *interval* between two events. If (x_1, y_1, z_1, t_1) are the coordinates of event 1 and (x_2, y_2, z_2, t_2) the coordinates of event 2, then the interval s (or four-dimensional 'distance') is given by

$$s^2 = c^2(t_1 - t_2)^2 - (x_1 - x_2)^2 - (y_1 - y_2)^2 - (z_1 - z_2)^2.$$

The minus signs in this equation are what emphasize the difference between time and space. They mean that s^2 can be positive or negative. It is also a fundamental requirement of relativity that the value of s^2 does not depend upon the coordinate system chosen to define (z, y, z, t). If s^2 is positive, the two events are said to be time-like separated and a signal can pass from the earlier to the later one (from 1 to 2 if t_2 is greater than t_1) so that the earlier event can influence the later one. If s^2 is negative, the two events are said to be space-like separated. No signal can pass between them; they are independent events without the power to influence each other.

Just as space and time can be amalgamated to give a four-vector (as we say) so can energy (E) and momentum (p_1, p_2, p_3). The corresponding 'interval' proves to be directly related to the rest mass m of the particle having this energy and momentum:

$$m^2c^4 = E^2 - c^2(p_1^2 + p_2^2 + p_3^2).$$

If the particle is at rest, its momentum is zero, and this relation then becomes just the familiar Einstein equation

$$E = mc^2.$$

VI

Exits and Entrances

Because of the uncertainty principle we are faced with choosing between alternative descriptions of the physical scene. Either we can know where particles are (in which case we do not know what they are doing) or we can know what they are doing (in which case we do not know where they are). If we make the former choice, we decide to describe things in terms of the coordinates of space-time. In the jargon of the subject this is called working in *configuration space*. If we make the latter choice, we decide to work with momenta (including, of course, the energy as one of the components of four-momentum). This is called working in *momentum space*. In either case we are taking a pretty comprehensive view—in configuration space our fields are defined at all spatial points and all times; in momentum space they are defined for all values of the momentum. So all-embracing a viewpoint came under criticism, for it greatly exceeds what we are actually able to observe.

The typical experiment in particle physics is a scattering experiment. We start with an initial state characterized by a target particle and a projectile particle incident upon it. When they come together an interaction takes place. We do not observe the details of this, all we see is the eventual final state consisting of the particles (including perhaps some additional ones created by the change of energy into matter) in the state in which the interaction leaves them. In other words, far from a configuration-space view of all points and all times we simply see the coming together of the initial particles before the collision and the separation of the final particles after it. What goes on in between is not observed. This gives a remarkable twist to our theatrical metaphor. The action on the stage is unseen, only the entrances and exits of the actors are noted from the wings!

The restricted nature of our knowledge is even clearer from a momentum-space point of view. Far from the four-vector momentum being able to take any value we like, it is restricted by the penultimate equation of Chapter V, which must hold with m the mass of the particular particle in question. This equation is expressed in words by saying that the particles 'must be on the *mass shell*'. All that this rather high-flown phrase means is that the momenta of particles must be subject to the constraints required by relativity if they are indeed to be particles of a specified mass m.

As they pondered these considerations some people felt impelled to whip out Occam's razor and cry 'Away with field theory with its physically unrealizable view of the whole of space and time. Instead let us restrict ourselves to a more modest attempt to calculate only what we can observe in a scattering experiment'! Such knowledge would amount to saying what sorts of final states are possible, and with what probabilities, when we are given a particular state from which to start. Thus all knowable information resides in these correlations between initial and final states of particles on the mass shell. There is a natural mathematical way of expressing such correlations. It is called a matrix. This is simply a rectangular array in which the rows and columns are labelled in a particular way. An example of a matrix would be a price table for selling bacon. Across the top will be indicated the various types, streaky, long back, etc., down the sides the various weights. If we want to find the cost of 12 oz. of streaky we run down the streaky column till we find the entry in the row corresponding to $\frac{3}{4}$ lb. The rows and columns of the particle physicist's matrix are labelled, respectively, by the initial and final states of the scattering experiment. From the corresponding entry we can calculate the probability of ending up with a specified final state if we started with the given initial state. (Technical note: the matrix entry is a complex number. The probability is given by the square of its modulus. This is how we always calculate probabilities in quantum mechanics.) This matrix is called the S-matrix. The S simply stands for scattering.

S-matrix theory

Heisenberg spent some time in the early 1940s working on S-matrix theory while his British and American colleagues were busy calculating the critical behaviour of lumps of uranium. However, the theory really began to develop in the late 1950s, mainly as a result of frustration at the inadequacy of trying to apply quantum field theory to describe strong interactions. The only techniques available were provided by the perturbation theory of Chapter V. That was all very well for quantum electrodynamics, where the expansion was in powers of a genuinely small quantity, $\alpha = \frac{1}{137}$, but we have already noted that the corresponding parameter for strong interactions is of the order of 1 or more, so that the sequence of terms shows no signs of getting smaller and hence making quantitative sense. It was certainly time to try another line of attack.

If S-matrix theory were to have any hope of success, it was clearly necessary to identify powerful properties which would impose conditions on the S-matrix so restrictive in character that they might actually serve to determine it, with only a small number of parameters left to be settled experimentally. Only then would a real theory be to hand. The first of these general requirements was rather easy to identify.

Given an intial state, the S-matrix determines the probabilities associated with possible final states. If we sum these probabilities over all the possible final states the answer must add up to one, since it is certain that *something* happens. This condition is called *unitarity*.

The second general property, called *crossing*, is more subtle. We can represent a scattering process, or its associated S-matrix entry, by drawing a bubble diagram like the figure. Inward directed lines will represent incoming particles, outward directed lines outgoing particles. Thus the figure represents a scattering

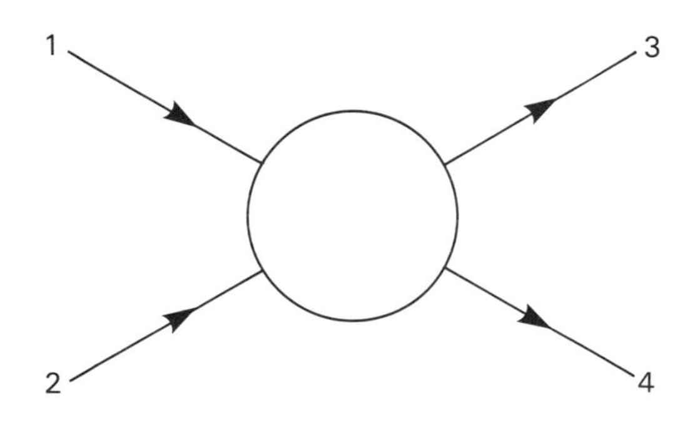

process in which

$$1+2 \rightarrow 3+4. \qquad \text{(A)}$$

Here we are reading the diagram in the obvious way from left to right. Suppose that instead we tried to read the diagram from bottom to top. If that is to make sense we shall have to reverse the arrows on 1 and 4 and it turns out that this means they will have to turn into their antiparticles, $\bar{1}$ and $\bar{4}$. If we do that, and read the figure in this different way, we find that we are thinking about a new process,

$$2+\bar{4} \rightarrow \bar{1}+3. \qquad \text{(B)}$$

Obviously this is obtained from the first process by crossing over 1 and 4; hence the name.

The change involved in reversing the arrow brings certain other consequences with it. Associated with process (A) there are two quantities. They are defined in terms of the four-momenta of the particles, but we need only concern ourselves with their physical characters. One, called s, determines the energy available for the interaction (A). (Technical note: $s=(p_1+p_2)^2$ in the four-vector sense so it differs from the s^2 of Chapter V.) The other, $t(=(p_1-p_3)^2)$, is the momentum transfer; that is, it measures the degree of deflection when particle 1 turns into particle 3. The conventions of special relativity (compare the end of Chapter V) require s to take positive values for the first process, and t negative values. When we reverse the arrows to produce the second process (B) these two variables interchange roles. The first (s) becomes the negative momentum transfer for the new interaction, whilst the second (t) becomes its positive energy. So far we have just been talking in a chatty sort of way. If you read the figure from left to right it looks like process (A), if you read it from bottom to top it looks like process (B). So what? Certainly both processes are possible things to think about and are expected to occur in nature. Without this discussion we would have supposed them unconnected with each other. The notion of crossing is to take this chat deadly seriously. The great idea is that there is a single entity, a single function as the mathematicians say, which describes both process (A) and process (B). It depends on the two variable quantities s and t. When s is positive, t negative, we have (A); when s is negative, t positive we have (B). Two processes for the price of one function!

Actually it is even more of a bargain than that for we can cross any pair of particles, producing in addition to (A) and (B) further processes like

$$2+\bar{3}\rightarrow\bar{1}+4. \qquad \text{(C)}$$

(It is a little more complicated to explain what this does to s and t, but never mind.)

The attractions of crossing extend beyond the many-birds-with-one-stone principle. Combined with unitarity it provides a tight constraint on the S-matrix. The same function has to provide probabilities adding up correctly for process (A) when s and t take values corresponding to that process, and for process (B) in its (s, t) region, and process (C) in its (s, t) region. It requires a pretty dexterous function to achieve this feat. In fact one might be surprised that such a thing is possible at all were it not for the fact that the Feynman integrals of the last chapter provide examples of it happening. This is perturbation theory acting in its second, qualitative, role of providing a 'theoretical laboratory', a source of insight into the structure of relativistic quantum mechanics. Without it we should have found it difficult to think of crossing.

However, crossing would have no meaning were it not for the third and most mysterious general principle, *analyticity.* A few mathematical ideas are unavoidable at this point. The first is that of a function. In its most general form a function is simply a recipe for producing one number given another. In mathematical notation you give me a number x and I give you another number $f(x)$. The rule may be something very simple, like squaring the number given: you say 1, I say 1; you say 2, I say 4; and so on. In this case the function is simply $f(x)=x^2$, that is, squaring x is the recipe. A question to which mathematicians address themselves is whether if you know a function for some range of values of x (all positive x, say, or x between 1 and 2, or whatever), can you say what its values will be for all the other values of x? Only certain particularly nice types of functions have this property; they are called analytic functions. Obviously x^2 is an example, the recipe works whatever value you choose to give to x. (Usually analytic functions are defined differently, in terms of what the mathematicians call complex variable theory, but this 'global' property, as it is sometimes called, is not only the easiest to explain but is also the one that is relevant for our present purpose.) If S-matrix elements were not analytic functions, crossing would be an empty statement. Unless, for example, knowing the function with s positive, t negative (region for process (A)) actually *determines* what values the function takes for s negative, t positive (region for process (B)), there is no content in saying that the two processes are described by the *same* function.

Requiring S-matrix elements to be analytic functions proves to be a step of enormous power. There is a potent store of mathematical machinery which can be brought to bear, often with formidable effect. The stirring together of these three properties of unitarity, crossing, and analyticity produced a heady brew and much employment for theoretical physicists. However, the upshot of this activity has proved less than was at one time hoped for it.

Critique

The development of S-matrix theory was characterized by a certain degree of sectarian strife. Some of us who worked in the subject were easy-going eclectics, content to try our hands at any approach to relativistic quantum mechanics which looked promising. Not so some of our senior and more inspiring colleagues. It was not so much a question of its being expedient to be on the mass-shell as of its being sinful to be anywhere else. In particular, they proclaimed the demise of quantum field theory. There is always a tendency for such dogmatism to generate heat as well as light. Writing more than ten years afterwards, the episode is seen to have a quaint period charm. Partly this is because recent years have seen, as Chapters VII and VIII attempt to sketch, a great resurgence of field theory. As with Mark Twain, the announcement of its death has proved exaggerated. Partly it is because the S-matrix endeavour looks a good deal less beguiling than it did in those brave early days. This is due both to problems of principle and also to difficulties of practice. We turn first to the former.

The first headache for S-matrix theory is to know where its basic principles come from. Unitarity is easy, since it is clear that whatever probabilities we calculate they must add up to one. But what about the other two, crossing and analyticity? There is some reason to think that the latter is connected with causality, the principle which says that the bell doesn't ring till the button is pressed, or (more relevant for a scattering experiment) that nothing comes out till something comes in. This expectation derives from the fact, long known to electrical engineers and the like, that applying the condition of causality to non-relativistic systems does produce analytic properties in the functions which describe them. However, the adaptation of this idea to the relativistic motion of massive particles encounters severe technical difficulties. As for crossing, we are even more at sea. The awful truth is that these two properties were first recognized in the Feynman integrals of quantum field theory, and they derive their plausibility from this fact. It seems not so easy to discard the old ways after all.

A second difficulty is that the scattering of our mass-shell particles is not quite the only class of possible observations. For example, the strong interactions have the effect of fuzzing out hadrons like the proton so that they no longer look like points and it is possible to measure the way that electric charge is distributed within them (more of this in Chapter VII when we talk about photon probes). Such determinations of form factors (as these distributions are called) represent experiments going outside a strictly S-matrix framework.

Finally there are difficulties of principle in applying the S-matrix approach to long-range forces like electro-magnetism. The neat separation into before and after, with the unknown scattering interaction in between, is not appropriate for such wide-ranging interactions from whose effects particles can never completely escape.

Added to these difficulties of principle were severe problems of actual technique. As the energy of a reaction increases, so the range of possible final states increases also, for additional energy means the possibility of more extra particles created in the final state by the materialization of some of this energy as matter, according to the Einstein transaction $E = mc^2$. In consequence the S-matrix gets more and more complicated and totally unwieldy to handle as the total energy goes up.

S-matrix theory played a valuable role in highlighting certain general properties (unitarity, crossing, analyticity) which are important aspects of relativistic quantum mechanics. This was a permanent gain which remains, despite the fact that almost all theorists now prefer to return to the lusher pastures of quantum field theory. We have seen in the last decade a transition from the austere generalities of the S-matrix to an ambitious and very promising attempt at the creation of a quite specific and detailed dynamical theory. This is quantum chromodynamics which we shall meet in the next chapter.

S-matrix theory also produced an idea which, though most would not now expect it to be a guiding principle of particle theory, is yet so elegant that it must be described. It is time to speak of the bootstrap.

Bootstrap

We started this book by saying that over the centuries there had been two basic types of theories of matter. One type thought of there being a single universal substance whose condensation or rarefaction produced the variety of observed matter. The other theory posited a small number of basic elements out of whose varying combinations the multiplicity of observed matter would be formed. It will be clear that the burden of our tale is a theory of the second type, with the different types of quarks playing the elemental role. S-matrix theory produced a startling third possibility, the bootstrap, which effectively said that everything was made out of everything else. According to how it pleased you to think of it, either there were no fundamental constituents, or everything played that role. The former is perhaps the better way to think of it. 'When everyone is somebody then no one's anybody.' The idea originated in the 1960s in the United States and it was propagated by Geoffrey Chew under the rousing slogan of 'nuclear democracy'. The notion of making things out of themselves had the air of breath-taking legerdemain about it for which a fitting metaphor was that of lifting oneself up by one's own bootstraps.

The basic idea can be illustrated by thinking of the ρ-meson (rho). This is a triplet of resonances, ρ^+, ρ^0, ρ^-, which are formed out of two pions. They clearly constitute a three-fold multiplet from the point of view of isotopic spin. We can picture the effect of the ρ in pion–pion scattering by a process like the figure, where two pions collide, stick together for a while to form the short-lived ρ-meson, and the ρ then decays to produce the two pions in the final state. This is the classic picture of an intermediate resonant state that we met in Chapter IV. Since this happens a lot of the time, producing the dominant effect on the

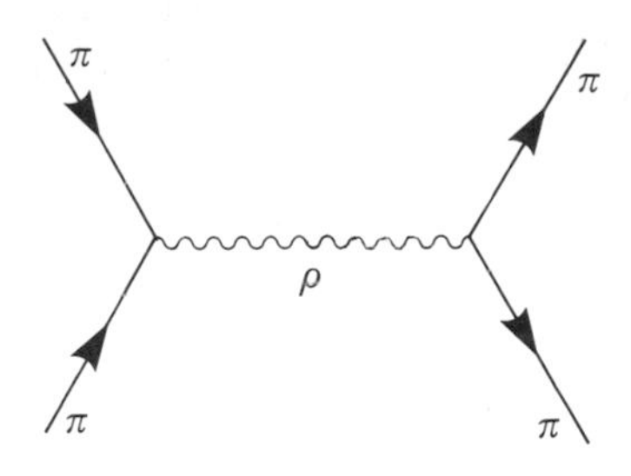

cross-section for the scattering of two pions in the relevant energy region, there must be quite a strong force which makes the pions likely to stick together in the ρ-state. Where does this force come from? Our picture of forces in relativistic quantum mechanics (Chapter 1) is that they are due to the exchange of particles. What particle would the πs like to exchange? One possible answer would be the ρ itself! This would give as the source of the force the second figure, which is obviously a

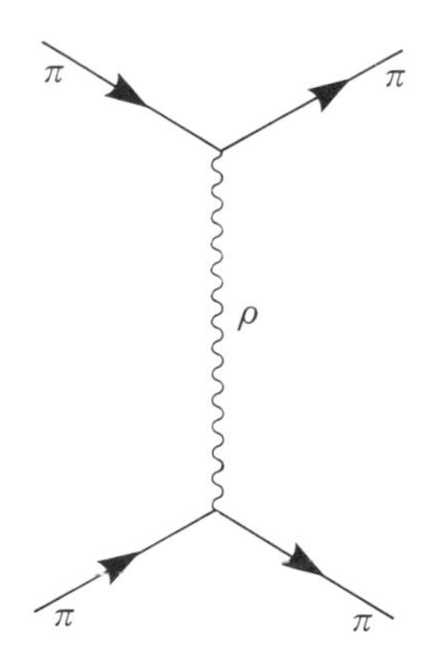

'crossed' version of the first. Our argument has curled round on itself in an intriguing way. To make the ρ we need a force between pions, that force being provided by the ρ itself. Clearly a delicate balance is required to make sure that this idea works in a self-consistent fashion. That the ρ provides just the right force to give a resonance at just the right energy to be itself will be a tight determining requirement on its properties. That is good. It puts predictive value into the bootstrap idea.

This example contains the germ of bootstrappery. It is the ρ which makes the ρ. Of course the pions themselves are still unexplained at this level of argument, but it is clearly conceivable that one could generalize these considerations to a wholly bootstrapped model in which everything was made out of everything else. The resulting theory would curl back on itself in all sorts of demanding self-consistency conditions. These might be so tight that there was just one solution. The physical world would be explained.

It was a grand idea. Unfortunately it has not come to much. Even at the level of getting the ρ it was difficult to make things work satisfactorily. The grand synthesis was infinitely more complicated and correspondingly inaccessible. Finally the nature of the microworld is unencouraging to egalitarian notions of nuclear democracy. There do seem to be objects which have a preferred primary role. It is time to return to the quarks.

VII

Quarks Reprise

'Particles behave as though they were made of quarks'. The simplest way of seeing whether this is really so would seem to be to take a look. In microworld terms this means deep inelastic scattering. It is the direct analogue of shining a torch to see what is there.

Deep inelastic scattering

To find out what is inside a proton we need a probe with which to poke around. It cannot be done with our fingers nor even with the most powerful electron microscope. Instead we shall have to use one of the fundamental interactions of nature. If we choose electromagnetic interactions the photon will be our tool. A simple way to make use of it will be to scatter electrons off protons, as in the figure, for the interaction takes place by the

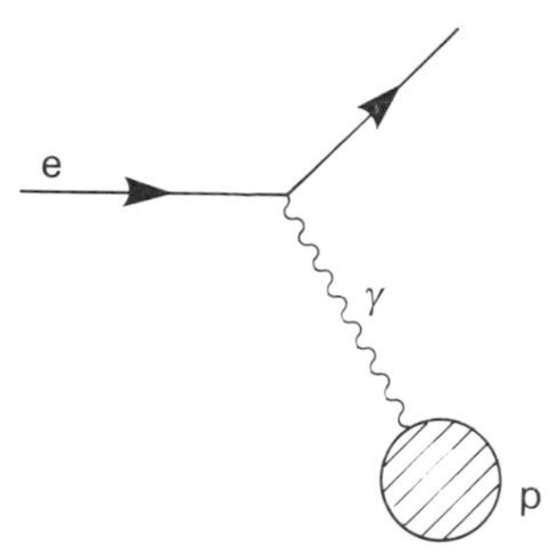

exchange of a photon. Because quantum electrodynamics is so well understood, the process of the emission of the photon by the electron is something which we know perfectly well how to calculate and can take for granted. This makes it possible in interpreting the experimental results to concentrate on the consequences of the proton's absorption of the photon. That is what we want, since it will tell us something about the structure of the proton. For useful information the scattering must be 'deep inelastic', that is at high energy and with large transfer of momentum from the electron to the proton. This is to make sure

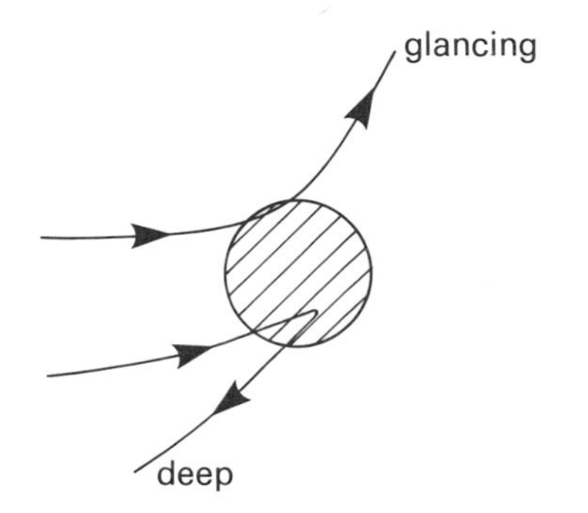

that the probing photon gets deep inside. A glancing incidence, with the electron little deflected, results in small penetration. The scatters which count are those where the electron is turned through a wide angle. If the photon carries across large momentum of amount Q it will probe distances within the proton whose scale d is given by an uncertainty-principle relation

$$d = \hbar / Q.$$

The proton may get broken up (that is the 'inelastic' part of the terminology) but that does not matter. The investigation is done by measuring only the final-state electron and not bothering to analyse the hadronic remains. Such experiments are consequently called *inclusive*, since they sum over all the possible sets of fragments left behind.

Deep inelastic scattering off protons was first performed using electrons from the two-mile linear accelator at Stanford. Great excitement was caused when a significant proportion of events with large deflections was observed. This indicated that one was seeing the presence of small point-like objects within the proton. The process could be compared to a rather bizarre experiment in which the nature of a christmas pudding (the proton) is investigated by firing bullets (photons) at it. If the pudding were just a cakey sort of mixture (if the proton were just a fuzzed out object) the bullets would tear through it little deflected (there would be only small transfers of momentum). However, if there were sixpences inside the pudding (if there are point-like constituents within the proton) then occasionally one of them would be hit and the bullets would be considerably deflected (deep inelastic scattering will take place off the constituents). This interpretation of deep inelastic scattering is called the *parton model,* parton being a distasteful neologism for point-like constituents.

The Stanford experiments were performed in the late 1960s. However, the idea of deep inelastic scattering as a probe of fundamental structure has an impressive pedigree stretching back far beyond that. In 1911 Rutherford scattered α-particles off atoms and found that there was quite a substantial amount of wide-angle deflection (deep inelastic scattering). He interpreted this as the presence of a positively charged object of point-like character (that is, judged by the standards of the day, which meant very small compared with the atomic dimension of 10^{-8} cm). He had discovered the nucleus.

It is possible to use probes other than that provided by the electromagnetic interaction. If we wish to make use of weak interactions the neutrino is the most suitable projectile. This is because it only has weak interactions, so that if it is scattered we know that they must be responsible. These experiments are more difficult to perform than those using electrons but they are nevertheless possible with effort and expenditure. Since particles interact differently with a weak interaction probe than they do with photons, the information obtained in this way is complementary to that given by electromagnetic interactions. The details of the modern theory of weak interactions will be discussed in the next chapter, but it will be clear that this extra information must help to determine the nature of the partons. Weak interactions provide a different perspective in observing the structure of particles, which when combined with the view afforded by the photon, gives a rounded picture of the whole.

When all the evidence from deep inelastic scattering of electrons and neutrinos is analysed a simple striking result emerges. The partons are found to behave exactly in the way that is expected of quarks! However, it is also found that the quarks do not account for all the matter that is contained within the proton. For example, they appear to carry only about half the momentum of the parent proton. The other half must be carried by constituents which do not interact with the electromagnetic or weak probes and are consequently invisible to us. It is supposed that they are the particles whose exchange produces the forces which make the quarks stick together in hadrons. Consequently they are called (oh dear!) *gluons*. We must return to them in more detail at the end of the chapter.

Confinement

The results of deep inelastic scattering intensify the mystery of quarks. Already from the patterns of hadronic spectroscopy (Chapter IV) we had learnt to take them seriously. Mesons appeared to be made of a quark and antiquark and baryons of three quarks. Now it seems that if we probe inside a hadron there are the quarks sitting there. Yet they do not turn up free and on their own. What happens in a wide-angle electron scatter is that the photon carrying the large momentum is absorbed by a quark within the proton. In consequence this quark is subjected to an almighty wham and we should expect it to be knocked completely out of the proton. (The bullets would certainly knock the sixpences out of the christmas pudding.) Apparently this does not happen. Curiouser and curiouser.

Sometimes it is necessary in physics to make of vice a virtue. As people brooded on quarks they came to feel that their elusiveness was of their essence, that they are characterized by only existing within the matter that is made of them, with no individual life of their own. It is always comforting to have a piece of terminology to describe a new and only partly understood idea. (It is the ancient notion that to name is to control.) We call the phenomenon *confinement*.

If quarks are confined, the level in the structure of matter that they represent is different in kind from all its predecessors. Atoms can be ionized, electrons torn off to reveal the nucleus within. That nucleus can be split up to manifest its component protons and neutrons. Such decompositions are the ultimate test of compositeness. Not so at this new level it seems. Tricks of intellectual perspective are very beguiling and there is a certain danger of hubris in the thought that it might have been given to this generation to discover the ultimate constituents of matter. Nevertheless something unprecedented is apparently occurring. It might be that we have reached the centre of the onion.

In an attempt to understand confinement better people have made toy models which mimic the effect without producing it from a fundamental theory. One such model pictures a hadron as a bag within which the quarks are able to rattle around a little. They respond freely to the photon probe of deep inelastic electron scattering, but then the strong walls of the bag prevent the struck quark actually flying off on its own. A related model pictures the quarks as if they were tied together by elastic strings. Normally these strings are rather slack so that again a quark can respond freely to the deep inelastic probe. As it flies off, however, the string tightens. However big the force that struck the quark, its effect will eventually be overcome by the pull of the string as it stretches and its tension increases. Of course this crude pictorialism is not to be taken literally, even when it is wrapped up in a more mathematical formulation. It is a crucial question whether a true theory of confinement would require some modification of basic physical principles or whether it is a property that could be possessed by certain types of relativistic quantum mechanical system. The answer is believed to be that it does not require any addition to the fundamental principles of Chapter V. The origin of that belief we now explain.

Gauge theories

Recent years have seen a great revival of interest in quantum field theory. This renaissance has been due to the recognition of the specially attractive properties possessed by a class of theories called gauge theories. They can be thought of as embodying the quintessence of symmetry. In order to try to understand them we turn again to the example of isotopic spin rotations, those operations in that curious fictitious space which can change a proton into a neutron, or a π^+ into a π^- (see Chapter IV).

The invariance of a theory under isotopic spin means that there are no preferred directions in the fictitious isotopic space. Putting it another way, if we set up axes in this isotopic space they can be oriented how we like; it makes no difference to strong interactions. Field theory combines the fictitious isotopic space with the points of real space-time at which the fields are evaluated. A way of picturing this combination is to think that at each point of space-time there is a set of axes in the fictitious isotopic space

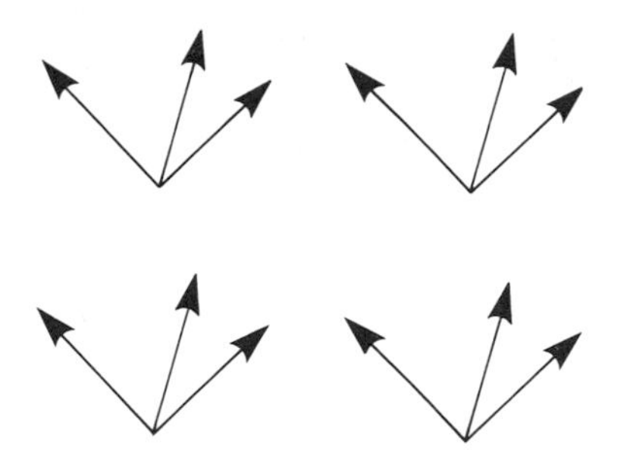

which we can use to define the isotopic spin properties of a particle located at that space-time point. Isotopic spin symmetry means that the orientation of these axes is arbitrary, but having chosen some orientation it has to be the same at all space-time points. This is shown in the figure. The sets of axes must always be mutually parallel. In concrete terms, if we make an isotopic rotation that turns a proton into a neutron this happens independently of where the particle in question happens to be. If there are several protons at different points they *all* turn into neutrons. This is exactly what invariance of the theory under isotopic spin means, no less, and equally no more.

However, eventually people began to wonder if there might be some special theories which have 'more'. Could it be that the isotopic axes could be rotated *independently* at the different space-time points, perhaps here turning a proton into a neutron and there not? In that case the first figure would be replaced by

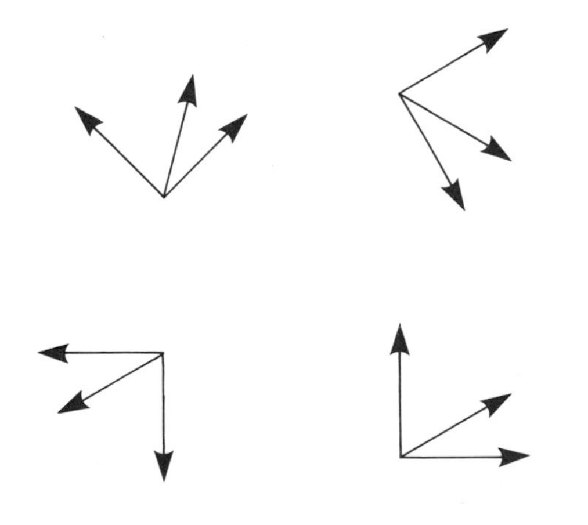

something like the second figure. It is not hard to believe that a theory with this remarkable property would be of a very specific kind and might prove to have other unusual features. Such theories do exist. They are called gauge theories. (The reason for the terminology is that a theory of related type is electromagnetic theory, in which the role of the transformations corresponding to the independent rotation of isotopic axes is played by what are called 'gauge transformations of the second kind'. Mathematically, electromagnetic theory is the simplest gauge theory but it does not lend itself so readily to the type of picture-book exposition which we are attempting for isotopic spin.) It turns out that an indispensable ingredient of gauge theories is that they contain interactions mediated by particles of spin 1. These particles are called gauge bosons. (The gauge boson of electromagnetic theory is just the photon.)

The gauge theory idea is of such importance that the reader must bear with a little recapitulation before we hasten on to its exploitation. Gauge theories are the *ne plus ultra* of symmetry. We thought we had a pretty symmetrical theory when we encountered the idea of isotopic spin. This implied that a neutron was as good as a proton, so that we could change one into the other without altering the strong interactions at all. However, if we did so we had to be thorough going. All the neutrons had to become protons. There could be no exceptions, with the neutrons here changing but those over there not doing so. In a gauge theory, on the other hand, things are so symmetrical that just such a position-dependent change is possible. This imposes a much stronger constraint on the structure of the theory. Perhaps the point can be illustrated by a trivial example. Suppose nine pennies are set out in a square, all with heads upwards. If we turn them all over we have an equally symmetrical array but with the simple change that they are now all tails. If the array is of equal symmetry when only some of the pennies are reversed (a position-dependent choice) then we can deduce the extra constraint that the pennies are double-headed, their tails being the same as their heads.

The application of the gauge principle to isotopic spin was proposed by two Americans, Yang and Mills in 1954, and these theories are generically called *Yang–Mills theories.* The same idea occurred simultaneously to a Cambridge research student contemporary of mine, Ron Shaw. He did not publish it (other than in his Ph.D. dissertation) because he believed that such theories would have a fatal flaw. The gauge particles would have to be massless. This is certainly, and acceptably, so for the photon. However, there are no other massless spin 1 particles around in nature, least of all ones with strong interactions with protons and neutrons!

Clearly if this conclusion were correct Yang–Mills theories would be valueless and I would not be writing this section. However, people eventually found a subtle way round which kept the admirable symmetry of gauge theories whilst giving the gauge particles mass as desired. One could get rid of the bath water but keep the baby. The essential idea was to recognize that it is possible to have a highly symmetrical theory but that the states it produces need not exhibit the same exacting degree of symmetrical perfection. In that way the gauge particles can elude the symmetry constraints which otherwise would force them to become massless.

This phenomenon is called *spontaneous symmetry breaking.* A familiar example is provided by the existence of permanently magnetic materials. The origin of this lies in the presence of atomic constituents which act like small magnets. Considered in isolation these small magnets are free to orient themselves as they please. There is a basic isotropy with no preferred direction. However, when an aggregate of such sub-systems is formed then its lowest-energy state is the one in which all the little magnets line up parallel to each other, as in the figure. This has the

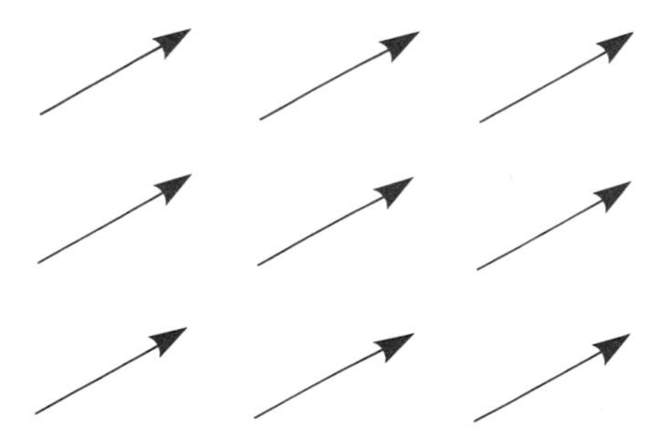

consequence that their magnetic effects add up, rather than cancelling each other out as they would for random orientations. The large-scale magnetic properties of the material are caused by this cumulative effect. There is now a preferred direction, namely that of the alignment. Thus the state (the magnet) has a lower symmetry (it is not isotropic but strongly directed) than the basic law (the behaviour of an individual atomic magnet).

Without spontaneous symmetry breaking the gauge idea would be restricted to the case of electromagnetism, where the symmetry is unbroken and the photon massless. In fact, the successful application of the idea outside electromagnetism also requires a notion of great importance, invented by Peter Higgs of Edinburgh, which unfortunately is too technical to explain here. (If I can indulge myself in a somewhat gnomic aside, Higgs' trick removes certain other unwanted massless objects, called *Goldstone bosons*, which would otherwise be there to plague gauge theories which had spontaneous symmetry breaking.)

Thus improved, gauge theories turn out to have many remarkable properties which stem from the high degree of symmetry they still retain, even in their spontaneously broken form. Ordinary quantum field theories have a rather large number of parameters associated with them which can only be determined by appeal to experiment. These parameters are the coupling constants associated with the basic processes that the field theory contains within itself. In Chapter IV we encountered the electric charge as the prime example of a coupling constant; it determined the strength of the interaction that the electron had with the photon. Every basic interaction of this particle with that will have its own coupling constant. A symmetry principle, which says that sets of particles behave in similar ways, will impose relations between these coupling constants. Remember the discussion of the charge independence of nuclear forces at the end of Chapter I. That symmetry (which is a particular case of isotopic spin symmetry) imposed relations between processes like a proton emitting a π^+ or π^0, and a neutron emitting a π^0 or π^-. The more all-embracing the symmetry, the more wide-ranging the relations will be and in consequence the smaller the arbitrariness left to be determined by experiment. Gauge theories are the most symmetrical quantum field theories. Therefore they are also characterized by the fewest adjustable parameters. They are tightly knit.

These relationships between couplings also enable things to happen which cannot be achieved in a more arbitrary situation. A striking example is the renormalizability (Chapter V) of gauge theories. It proved very difficult to find field theories involving spin-1 particles of a non-vanishing mass which were renormalizable. (Quantum electrodynamics is all right because its spin-1 boson is the zero mass photon.) In fact only gauge theories appear to succeed in having that desirable property.

By now the reader will no doubt have joined most contemporary particle physicists in the ranks of those who think that gauge theories are 'A Good Thing'. Let us pause to take stock.

A gauge theory is a very highly symmetrical and consequently very tightly knit version of quantum field theory. It incorporates the idea that symmetry operations can be performed independently at the different points of space-time. (We illustrated that by talking about isotopic spin, but the idea can readily be extended to the symmetry associated with any Lic group. In particular there is a gauge theory associated with SU(3).) A gauge theory must contain within itself gauge bosons which have spin 1. By using the idea of spontaneous symmetry breaking, these gauge bosons can avoid the embarrassment of being forced to be massless. The gauge symmetry gives the theory many beautiful and special properties. One is renormalizability. Another must now attract our attention.

Asymptotic freedom

A very important property possessed by gauge theories goes by the stirring name of asymptotic freedom. The strength with which an interaction manifests itself depends on the length scale on which it is being sampled. This is a consequence of the renormalization effects of Chapter V. We saw there that the fact that the vacuum could act as a polarizable medium meant that effective charges got modified. The discussion was in terms of quantum electrodynamics but similar ideas apply to the coupling strengths of any interaction. These effects of vacuum polarization can clearly be expected to depend on the size of the sample region. We have so far tacitly ignored this, as did particle theorists for many years. The time has come for us, as it did for them, to look at what this implies. It has been one of the recent triumphs of quantum field theory that techniques have been developed which are capable of answering such enquiries as how various interactions appear when viewed on a very short or very long length scale. Because of the reciprocal relation between space and momentum, an alternative way of putting the same question is to ask how interactions behave at large or small momenta.

In gauge theories the coupling strengths are found to become progressively weaker as distances decrease. The particles behave more and more as though they were free of interactions. They are 'asymptotically free'. This is an exciting result. It provides a natural explanation of what is found in the short-distance exploration of deep inelastic scattering: the quarks free to rattle in their bag or with their connecting strings lying slack. More exciting still is the fact that theories whose couplings weaken at short distance have the converse property that these couplings strengthen at large distances. Here is the making of an explanation of the strong walls of the bag, the tightening extended strings. Such an effect deserves a slogan: '*infrared slavery*'. 'Infrared' because this means long wavelength or large distance; 'slavery' because the increasing interaction forges chains to restrain the would-be escaping quarks. In a word, we have confinement.

Gluons

The gauge theory which produces confinement must be that which describes the interactions of quarks within hadrons. That is, it must deal with the forces between the quarks which are produced by the exchange of gluons. The 'strings' which tie the quarks together are, therefore, tubes of force of the gluon field, similar to the tubes of force of thc electromagnetic field which nineteenth century physicists were so keen on visualizing as a picture of electric forces between charges. It is time to take a closer look at these gluons. They are connected with a new aspect of the quarks.

Despite its great success in explaining hadron spectroscopy, the quark model had long been recognized as having a puzzling feature. Consider the isotopic quartet of Δ particles, and in particular its doubly-charged Δ^{++} member. Δ^{++} must be composed of three u-quarks in order to get total charge 2. Since it has spin angular momentum $\frac{3}{2}$ it turns out that the spin state of the three spin-$\frac{1}{2}$ u-quarks is required to be symmetric. The Δ^{++} is also the lowest energy state of 3 u-quarks known and that leads us to expect, on the basis of vast experience of such calculations, that they will be in a state which is spatially symmetric. Putting those two facts together means that the three u-quarks are in a state which is symmetrical in both space and spin, which means that it is altogether symmetric. I have to ask you to believe that this implies that the u-quarks appear to be behaving as though they were bosons (that is, they like to be all symmetrically in the same state). Yet they are quarks with spin $\frac{1}{2}$ and so should be fermions (and no two should ever be in the same state). What has Pauli got to say about that? The relation between spin-and-statistics is too deeply entrenched in relativistic quantum mechanics for it to be lightly abandoned. Tentatively people began to suggest that maybe there was some extra property of the quarks which meant that the three u-quarks in the Δ^{++} need not all be identical after all. With the relentless facetiousness so characteristic of the subject, they called the hypothetical distinguishing attribute *colour* (or, more literally, since they were Americans, 'color'). The idea would be, say, that there were red, yellow, and blue quarks and that the Δ^{++} was made up of one of each. The u-quarks would then no longer be identical and so Fermi statistics would not operate and their state would be allowed to be symmetric.

At the time this seemed a rash speculative steam-hammer to crack what might be just a tough dynamical nut. (Maybe the state was not spatially symmetric after all.) However, it was later realized that there were experiments which effectively counted the number of quarks and that these would not give the right answer unless there was the tripling of quark types implied by colour. One process of this kind is the decay of π^0 into two photons, which is pictured as going through a quark–antiquark intermediate state. The more $q\bar{q}$ states there are accessible, the faster the reaction will go (it is like a phase-space effect; see p. 46) and the factor of 9 resulting from the colour tripling is just what is needed to bring calculation and measurement into agreement. Another process of great importance for quark counting is the high-energy annihilation of electron and positron. The pure blob of energy thus created can materialize into any available state of matter. The most efficient way for it to do so is as a point-like particle–antiparticle pair. If the pair is e^+e^- again, or $\mu^+\mu^-$, the final state consists of leptons. However, if the final state consists of hadrons, the pair must have been a pair of point-like hadronic objects, in other words a quark–antiquark pair. (Confinement will not allow the quark or antiquark itself actually to appear; they have subsequently to turn themselves into acceptable non-confined particles.) Thus the cross-section for

$$e^+ + e^- \rightarrow \text{hadrons}$$

at high energy is connected with the number of possible $q\bar{q}$ states which can be formed. Each contributes an amount proportional to the square of that particular quark's charge. To get the counting right it turns out again that there have to be three times as many $q\bar{q}$ pairs as we would have thought without colour. Thus the speculation is by no means as rash as it first seemed.

The invention of colour opens up important possibilities for particle physics. With all quarks becoming threefold a new SU(3) group enters the picture. This is the SU(3) of colour which has transformations doing such things as changing blue quarks into yellow quarks or yellow into red. It is very important to distinguish this new SU(3) of colour from the old SU(3) of Chapter IV. The latter concerned isotopic spin and strangeness and so had transformations which did things such as changing a u-quark into a d, or a d-quark into an s. In the current jargon quantities like isotopic spin or strangeness are called flavours, to distinguish them from colour. The new SU(3) of colour leaves flavour untouched. A u-quark remains a u-quark, but it changes from red to yellow. An important feature of this new SU(3) is that it is exact. Red, yellow, and blue quarks all have identical masses.

The gauge idea can be applied to the colour SU(3). The corresponding gauge particles are the gluons. There is no need to invoke spontaneous symmetry breaking. The gluons are massless, but no matter, for they will also be confined. This is because they too must be coloured and any object with colour will have colour tubes of force attached to it which will serve as confining strings. The only unconfined objects capable of free individual existence are those of zero colour. Such 'white' combinations of red, yellow, and blue turn out to be just the baryons (qqq) and mesons ($q\bar{q}$) with which we are familiar.

The colour gauge group goes beyond a qualitative understanding of the phenomenon of confinement. It provides a theory—called *quantum chromodynamics*—which can be the basis of quantitative calculation. So far the applications of QCD (as it is acronymically referred to) have been in high energy regimes, such as deep inelastic scattering, where the asymptotic freedom of the theory enables us to use perturbation theory techniques as the gluon couplings become small. Results are encouraging.

The discussion of this chapter contrasts sharply with that of Chapter VI. Instead of the generalities of the S-matrix we have the particularities of a specific dynamic theory, QCD. However, it may just be worth reminding ourselves that so far no *direct* evidence of gluons' existence has been obtained. In contrast with quarks they are flavourless, and so cannot be tasted by the probes of deep inelastic scattering.

VIII

Denouement

The last few years have seen dramatic developments in particle physics. One of the most important has been the drawing together of weak and electromagnetic interactions.

The unification of weak and electromagnetic interactions

At first sight weak and electromagnetic interactions look like chalk and cheese. Not only is there the question of their contrasting strengths, but there are also the other properties by which they differ. Electromagnetism conserves parity and strangeness, weak interactions respect neither. An unpromising marriage for such ill-matched partners, one might think. Much the same sort of thoughts must have been in the minds of nineteenth century scientists when they contemplated the apparently very different phenomena of electricity and magnetism. Yet the experimental results of Faraday and the deep theoretical insight of Maxwell led to their unification in the theory of electromagnetism which is the most splendid triumph of nineteenth century physics.

The first problem to solve in trying to bring together weak and electromagnetic interactions is the contrast in the strengths of the interactions (see Chapter I). The way to an answer lies in the hypothesis of the *intermediate vector boson.* Consider the weak interaction corresponding to the decay of a μ-meson into an electron, neutrino, and antineutrino:

$$\mu^- \rightarrow e^- + \nu + \bar{\nu}.$$

The idea will be to picture this as a two-step process, the μ emits a boson 'W^-' turning itself into the neutrino and the 'W^-' then turns itself into the electron and antineutrino:

$$\mu^- \rightarrow \text{`W}^-\text{'} + \nu$$
$$\hookrightarrow e^- + \bar{\nu}.$$

We have written 'W^-' in quotation marks because it is a particle only in a Pickwickian sense. Technically we express this by saying that the 'W^-' is a virtual particle. What that means is something like this. At first sight the two-step process looks rather like a resonance effect (Chapter IV) with the 'W^-' a short-lived intermediate state. This would indeed be so if the energy provided by the muon were large enough actually to create the 'W^-'. However, we shall see shortly that intermediate vector bosons are believed to be very heavy particles, so that the 'W^-' cannot be literally there as a halfway house in the process of μ-decay. Nevertheless, that fact does not prevent it having an influence on the process. We have said that forces in particle physics are pictured as due to the exchange of particles (Chapter I). If these particles have only an intermediary role, and so do not appear in the initial or final states, they are not required to be real particles 'on the mass shell' (Chapter VI). Instead they can be virtual particles (off the mass shell), making a significant contribution even in regimes where there is not enough energy to create a real (on the mass shell) particle. The formalism of Feynman integrals (Chapter V) is based on just such a virtual particle picture of field theory interactions. It is likely that this discussion will appear to have an air either of mystery or of sleight of hand. Once again it is necessary to appeal for the trust of the reader. Strange as these ideas may seem when stumblingly expressed in allusive prose, they correspond to a theory which is mathematically well defined.

The possibility of reconciling the strengths of weak and electromagnetic interactions turns on the mass of the W. If the W coupled to leptons with a coupling constant g (which would be the analogue of the electric charge e) then it turns out that the probability for μ-decay would be proportional to

$$\left(g\frac{m_\mu}{M_W}\right)^4,$$

where m_μ is the μ-mass and M_W the W-mass. Thus g could be quite large (comparable with the electric charge) but the μ-decay could still be weak because M_W was very large. In a general weak interaction process one would find a similar factor with m_μ replaced by E, the energy available in the interaction,

$$\left(g\frac{E}{M_W}\right)^4.$$

Thus weak interactions would be suppressed with respect to electromagnetic until one attained energies sufficiently high for the intermediate vector bosons to be created as real particles (E of the size of M_W). Then weak and electromagnetic interactions would have become of comparable strength and their hidden unity (assuming it to be there) would then be made manifest. To get the numbers right for ordinary weak interactions like μ-decay it is necessary to assume that the W is very heavy, 80 to 100 times the mass of the proton. Unfortunately this means that the production of a W would require more energy than is at present available in the laboratory. Therefore the acid experimental test of actually discovering an intermediate vector boson eludes our grasp at the present time, though it could be within the capacity of the SPS at CERN when it is used as an antiproton–proton collider in the early 1980s (Chapter II).

The W thus emerges as a particle which mediates the weak interactions. Study of the angular properties of μ-decay shows that if it proceeds via a W then the W must have spin 1. That is what is meant by calling it a vector boson; vector means spin 1 in our jargon. Of course this is essential if there is to be any comparison with the photon of electromagnetism which also has spin 1. It also makes one think of gauge theories. Pursuing that notion led Steven Weinberg of Harvard and Abdus Salam of Imperial College, London, to make independently in the late 1960s the suggestion of a specific theory combining weak and electromagnetic interactions. The first problem such a theory has to tackle is how to make the electromagnetic interactions reflection invariant whilst permitting the weak interactions not to conserve parity (Chapter III).

Parity violation is built into the theory by separating out right- and left-handed spin states of the spin-$\frac{1}{2}$ particles. They can either have their spin angular momentum making a right-handed rotation about their direction of motion ((a) in the figure, which corresponds to a right-handed corkscrew sense) or a left-handed

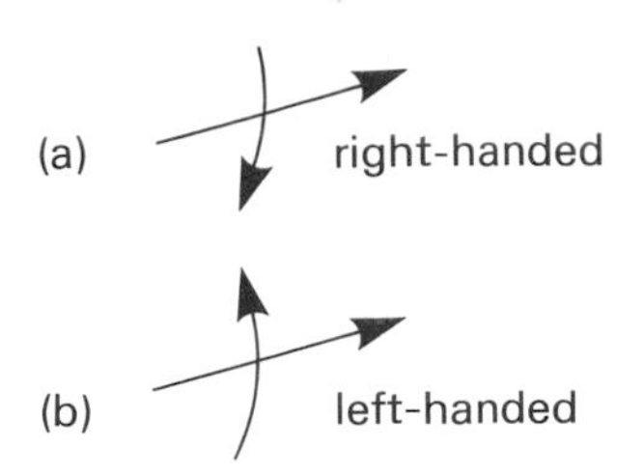

rotation (b). Treating these two states differently is manifestly a reflection-non-invariant thing to do. In fact the difference is extreme for neutrinos which occur in nature only in the left-handed form. Such a pure separation into exclusive left-handedness is only possible for massless particles, so that electrons and all the other particles with which we are concerned will have both left-handed and right-handed states. However, the Salam–Weinberg theory will still treat the two states differently. For example, the left-handed electron will be paired with the left-handed neutrino to form a doublet under the gauge group, but the right-handed electron will have to form a singlet all by itself. Clearly the theory will have to be cleverly constructed so that this asymmetry between right and left does not affect that part of the interaction to be identified with electromagnetism. Because of its parity conservation the latter must be even-handed.

It turns out that a gauge theory of weak and electromagnetic interactions can be constructed with all the desired properties. It is called SU(2)×U(1) because in fact it conflates a pair of gauge groups, yet another SU(2) group and a second group called U(1). (They are combined by what the mathematicians call a direct product.) This putting together of two groups rather than using just one is a rather unaesthetic feature which it would be nice to avoid, but it appears to be the way things are. It has the consequence that an additional experimentally determined parameter is required to fix the theory, which is not as tightly knit as it would be with a single gauge group (Chapter VII). Essentially this parameter measures the balance between the two groups, SU(2) and U(1). It is called the *Weinberg angle*, θ_W.

There are four gauge particles of spin 1 associated with SU(2)×U(1). One of them is the photon. Two others are the charged intermediate vector bosons W^+ and W^- which we have met already. (W^+ is just the antiparticle of W^- and plays an analogous role in μ^+-decay to that played by W^- in μ^--decay.) The fourth particle is a heavy neutral intermediate vector boson Z^0. The theory cannot do without Z^0 and in consequence a new type of weak interaction is predicted. Its nature is most readily understood by considering neutrino scattering.

A neutrino scatters via weak interactions only. The effect is mediated by the exchange of an intermediate vector boson, emitted by the neutrino and absorbed by the target. The emission process is a crossed version of boson emission in μ-decay and takes the form

$$\nu \rightarrow \mu^- + W^+.$$

The neutrino emits a W^+ and to conserve charge has to turn itself into a negative muon. (There is also a related process in which a ν turns into an electron.) When the W^+ is absorbed by a quark in a neutron it will turn the latter into a proton. The process is called in traditional language a *charged current* process. 'Current' is used in analogy with photons coupling to electric currents, whilst the epithet 'charged' draws attention to the fact that the W^+, in contrast to the neutral photon, carries electric charge. It therefore leaves the incident neutrino converted into a charged lepton, the μ^-. Reactions of charged current type are comparatively easy to investigate because the charged lepton in the final state is a readily detected signal that the interaction has occurred.

Given the existence of the Z^0 there must be a similar process

$$\nu \rightarrow \nu + Z^0,$$

in which the ν emits a neutral Z^0 and so, for reasons of charge conservation, must itself remain unchanged. The Z^0 is subsequently absorbed by the target. These reactions are appropriately enough called *neutral currents.* They are much more difficult to detect than the charged currents since the elusive neutrino in the final state does not show up in the way a charged particle would. In fact when Salam and Weinberg proposed their theory, neutral current reactions had never knowingly been seen. (As is so often the case, there were a few 'odd-ball' events which experimentalists had not been able to make sense of in the absence of theoretical guidance of what to look for.) The prediction of this new weak interaction, involving neutrinos scattering without becoming muons and mediated by the exchange of a Z^0, was an inescapable consequence of the SU(2)×U(1) theory. The occurrence of neutral weak currents was therefore the first acid test the Salam–Weinberg theory faced. It passed with flying colours when the neutral current was discovered at CERN. Moreover, the detailed investigations of these phenomena which followed have shown that their form is just that indicated by the theory with a value of the Weinberg angle θ_W of about 30°.

The prediction of neutral currents was the first triumph of the unified theory. More successes were in store as a consequence of thinking about how quarks fitted into the scheme of things. Before discussing that it will be useful to come clean about neutrinos. There are more of them than we have so far acknowledged. In fact there is one type of neutrino, ν_e, which goes with the electron and another type, ν_μ, which goes with the μ. This was shown by taking neutrinos resulting from the decay of fast π^+s and studying their charged current interactions. The neutrinos from pion decay are overwhelmingly ν_μs, since the dominant reaction is

$$\pi^+ \rightarrow \mu^+ + \nu_\mu.$$

(A similar decay with μ^+ replaced by e^+ is ten thousand times rarer for reasons which are well understood but which we need not attempt to describe.) When these neutrinos are allowed to scatter it is found that in their charged current interactions they always turn themselves into muons and never into electrons. Now if there were only one type of neutrino, freely associated with either muons or electrons, then when it scattered in a charged current way we would expect that half the time it would turn into a muon and half the time into an electron. (The effect which produces an asymmetry between μs and electrons in pion decay can be shown not to be relevant here.) The fact that this is not the case can only be explained by the hypothesis of two types of neutrino, the ν_μ produced with muons (as in π-decay) and only capable of turning back into a muon, and the ν_e, inexorably associated in a similar way with the electron. We can picture there being two qualities, muonishness (or muon number, as we actually say) associated with the μ and ν_μ, and electronishness (electron number) associated with the electron and ν_e, both separately conserved. Thus the μ-decay reaction gets rewritten

$$\mu^- \rightarrow \nu_\mu + e^- + \bar{\nu}_e,$$

where ν_μ carries on the muon character of the μ^- whilst the particle e^- and antiparticle $\bar{\nu}_e$ have cancelling electron number. It appears that the mysterious doubling in nature that we first encountered in Chapter I, with the μ an almost redundant big brother of the electron, is characteristic of neutrinos also. There is more of the same to come.

We have travelled a great intellectual distance since the start of this chapter. Electromagnetism and weak interactions, which seemed very different fundamental forces, have now proved to be partners in a remarkable union. The contrasts in their apparent strengths are really just contrasts in the masses of the particles mediating them, the heavy (and let us recall, still hypothetical) intermediate vector bosons, compared with the massless photons. The crucial tests of the idea will come at energies large enough to produce Ws and Zs. Then electromagnetic and weak interactions should be of equal importance. Although the investigation of that regime awaits us in the future, the theory has also definite predictions for lower and more accessible energies. These are very successful and give us great encouragement to accept the theory's validity. In particular, there is the striking prediction of the existence of neutral current interactions.

Salam–Weinberg theory makes essential use of the powerful machinery of gauge theories. It introduces parity breaking by separating spin states into left- and right-handed versions but does so in a sufficiently clever way as to preserve the reflection invariance of the photon's interactions. The neutrino only exists left-handedly. (The antineutrino, by the way, is right-handed.) It also exists in two distinct versions, ν_μ associated with the muon and the ν_e associated with the electron. From the point of view of Salam and Weinberg, therefore, the leptons are organized in the following way. There is electronic matter, consisting of a doublet with the left-handed ν_e and the left-handed part of the electron, plus a singlet which is the right-handed electron. There is an exactly similar pattern of muonic matter:

$$\begin{matrix} \left.\begin{matrix} e^{\text{left}} \\ \nu_e^{\text{left}} \end{matrix}\right\} & \left.\begin{matrix} \mu^{\text{left}} \\ \nu_\mu^{\text{left}} \end{matrix}\right\} \\ e^{\text{right}} & \mu^{\text{right}} \end{matrix} \qquad \text{(A)}$$

We shall discover a striking similarity in the way the quarks line up.

GIM

To incorporate hadrons into the Salam–Weinberg scheme, it is natural to organize the quarks into left-handed doublets and right-handed singlets in analogy with the leptonic pattern. In the doublet there will have to be a difference of charge of 1 between the analogues of the e(μ) and the ν, so that the emission or absorption of $W^{\pm}$ can occur with the conservation of charge. With the quarks at our disposal u (charge $\frac{2}{3}$) is a natural choice for one member of a doublet but either d or s (both charge $-\frac{1}{3}$) could be the second member. In fact it is desirable to make use of both possibilities together, for one (using d) will give strangeness-conserving hadronic weak interactions (like the π-decay of Chapter I), whilst the other (using s) will give the strangeness-changing weak interactions which are characteristic of the decays of strange particles (Chapter IV). At this point the fact that the gauge idea introduces neutral currents poses a problem. No neutral current effects are known which change strangeness. If they were present they would manifest themselves in quite striking ways in particle decays, so that their absence is something of a facer for the theory. To illustrate the point, consider K-mesons. They are found to decay quite frequently in a variety of processes involving leptons, such as

$$K^+ \rightarrow \mu^+ + \nu_\mu,$$
$$K^+ \rightarrow \pi^0 + \mu^+ + \nu_\mu.$$

The presence of the combination of a charged μ with a neutral ν_μ shows that these are charged current effects. The corresponding reaction

$$K^+ \rightarrow \pi^+ + \nu_\mu + \bar{\nu}_\mu,$$

if it happened, would be a neutral current decay, for the leptons here are both neutral. This decay is totally unknown in nature. The answer to the difficulty lies in the GIM mechanism.

GIM is an acronym for an international consortium of theoretical physicists: the American Glashow, the Greek Iliopoulos, and the Italian Maiani. They produced a clever device for cancelling out the unwanted strangeness-changing neutral currents. For this a price has to be paid. It is the postulation of a new quark, c, of charge $\frac{2}{3}$, carrying a new flavour to which was given the (awful) name of charm. The GIM mechanism parallels the lepton structure (A) with a quark structure

$$\left.\begin{matrix} u^{\text{left}} \\ d'^{\text{left}} \end{matrix}\right\} \quad \left.\begin{matrix} c^{\text{left}} \\ s'^{\text{left}} \end{matrix}\right\} \qquad \text{(B)}$$
$$\begin{matrix} u^{\text{right}} & c^{\text{right}} \\ d'^{\text{right}} & s'^{\text{right}} \end{matrix}$$

Two comments are necessary. First, more singlets are required in (B) than in (A) since the quarks are all massive and no right-handed parts can be omitted as they are for neutrinos. Secondly, d′ and s′ are not the familar d and s quarks but linear combinations of them. That is to say d′ is a 'quark' which is mostly d but sometimes s, and s′ is similarly mostly s but sometimes d. This is essential to give the correct charged currents (both strangeness-changing and strangeness-conserving) whilst achieving the desired cancellation of the unwanted strangeness-changing neutral currents. The idea of such fickle objects as d′ and s′, not quite one thing nor the other, sounds strange but it is in fact a natural option within a quantum mechanical framework. The statement about the properties of d′ and s′ just corresponds to a statement about probabilities; d′ is most probably found to be d but has a small probability of being s, and vice versa for s′.

The GIM mechanism was recognized as a clever trick but many of us thought it a bit preposterous to invent a new quark and a new flavour simply to solve the problem. Clichés about steam hammers and nuts abounded. The laugh was with GIM when the quality of charm finally made itself manifest.

Before going on to describe how that happened, it might be as well to speak to a thought which may be growing in the reader's mind. We have encountered at least two examples (colour and GIM) where a speculative notion of great importance to particle physics has been greeted with some scepticism when first propounded. Does this just show what crassly conservative people the majority of particle physicists are? Well perhaps it does, but a plea in mitigation of that view is possible. Because of its fundamental character, particle physics attracts to itself a lot of highly talented people. Their agile minds are continually engaged in speculative leaps of one kind or another. Only a very few of these imaginative *tours de force* achieve, by luck or by deep intuition, the endorsement by nature necessary for them to survive and find a place in history (and a book of this kind). However quickly these other notions may fly forgotten as dream, their butterfly existence while it lasts makes it the more difficult to sort grain from chaff. (Somehow a mixture of metaphors seems appropriate to describe the process.) But now it may seem that the defence is more damaging than the charge it was meant to counter. Is particle physics theory then nothing but a plethora of febrile speculation, or a lottery with the odd lucky man drawing a winning ticket? That too would give a wrong impression of the ethos of the subject. It is a curious mixture of the conjectural and the methodical. In fact particle physicists believe themselves to be restrained in their speculations when compared with workers in some other basic sciences, who show a pronounced tendency to invoke fundamental changes in the laws of nature in order to explain phenomena which, with greater effort, prove to yield to more conventional understanding. It was of astrophysicists that the great Russian theoretical physicist, L. D. Landau, said that they are 'often in error but never in doubt!'

Charm

In early November 1974, by a strange coincidence, two experimental groups, one on the east coast of the United States and the other on the west, both came to the conclusion that they had discovered a remarkable new phenomenon. They had come upon a very narrow resonance with a mass about three times that of the proton. Sam Ting's group at MIT had produced it in proton–nucleus collisions, an experiment of great delicacy and cunning in which they had been engaged for some time. The result was so remarkable that they were aware of it for a while before they finally became convinced that it was a real effect and not just some quirk in the experimental system. They called this heavy (but by strong interaction standards long-lived) particle, J. Burt Richter's group at Stanford made the same discovery in a totally different way. They were using SPEAR in which two high energy beams, one of electrons, the other of positrons, are made to collide. The resulting annihilation creates a blob of pure energy (technically, a virtual photon) which can materialize in a variety of forms of matter. Machines using e^+e^- annihilation are coming to play an increasingly important role in particle physics because they have this ability to produce energy which can then materialize in an unprejudiced way into any accessible state of matter. Such machines are good for surveying what is going on in a high energy regime. They are open to the unexpected. SPEAR certainly fulfilled that role when on being tuned to the appropriate energy it showed the existence of this new very narrow resonance. The west coast people call it ψ (psi).

The J/ψ came as a shock because of its narrowness. This meant (Chapter IV) that it was a particle which lived a thousand times longer than the 10^{-23} seconds typical of hadronic resonances. What made it so reluctant to decay? The theoretical community was full of ingenious suggestions but when the dust settled it was clear that one explanation held the field. It was the presence within the J/ψ of a new quantum number, charm, a sort of elder brother to strangeness. The way it worked was peculiarly subtle. The J/ψ is a particle with zero charm but it is made up of a charmed quark c and a charmed antiquark $\bar{c}$ which stay far enough apart within the J/ψ not to annihilate each other straight away. Of course their equal and opposite charms cancel out, leaving $J/\psi = c\bar{c}$ charmless. In a word, it has *hidden* charm. But why then should this stop the admittedly charmless J/ψ decaying very fast into ordinary charmless particles like pions? The frank answer is that we don't know, but we have seen this sort of thing happening before. There is a meson resonance called ϕ (phi), which is a state of hidden strangeness made out of s and $\bar{s}$. However, it does not care to decay into pions of zero strangeness but much prefers to decay into a K-meson and a $\bar{K}$-meson, that is into a strange particle–antiparticle pair. The cancelling strangeness within the $s\bar{s}$ system of the ϕ seems to persist into the cancelling strangeness of the $K\bar{K}$ final state, rather than being totally swallowed up into non-strange pions. Presumably the same is true, *mutatis mutandis*, for J/ψ. It would very much like to decay into a charmed particle–antiparticle pair. However, particles with non-zero charm turn out to be rather heavy and the J/ψ does not have enough energy available to it to make a pair of them. It is like a man with £1.90 in his pocket who believes it is vulgar to tip with less than a note and who consequently cannot reward both the cook and the butler. This impasse holds the J/ψ impotent to decay until it eventually swallows its pride and turns into non-charmed particles. (The rule is not absolute; the ϕ, for instance, very occasionally turns into pions.) Why there is this strong desire for systems to manifest their hidden quantum numbers in their decays is not really understood. Once again a name is comforting. The effect is attributed to *Zweig's rule.*

The existence of charm could not be taken as fully established until particles of non-zero charm were also found. The investigation of this regime was undertaken at the e^+e^- machines, SPEAR at Stanford and a similar machine DORIS in Hamburg, immediately following the discovery of the J/ψ. Another narrow resonance, the ψ', was found which is simply a more massive relation of the ψ. As the energy increased this was followed by a sequence of ordinary broad resonances, indicating that Zweig's rule was no longer having an inhibiting effect but the particles of hidden charm, which these resonances represented, were now sufficiently massive to decay into actual charmed particles. An intensive search of this latter region then yielded the desired direct evidence of these charmed particles' existence. Charm had come to stay.

Upsilon and Tau

Wonders never cease it seems. The next time it was the turn of the central United States to provide the excitement. In an experiment at Fermilab just outside Chicago, Leon Ledermann discovered in 1977 yet another very narrow resonance. The experiment was in essence a higher energy version of Ting's. The new particle, christened Υ (upsilon), is more than nine times more massive than the proton, or three times heavier than the J/ψ. By now the theorists knew how to explain it. Why, it must be yet another type of quark manifested in a hidden mode. I regret to have to tell you that its characteristic flavour is the quality of bottomness or, to a dissident and diminishing minority, beauty. Anyway, it is a b quark. Everyone believes that there is another yet heavier quark around, denoted by t and endowed with topness (or truth).

The reason for this belief is the expectation that the pattern (B) of p. 114 a lepton column like (A) of p. 112 recurs, so that a new column is formed:

$$\left.\begin{matrix} t^{\text{left}} \\ b^{\text{left}} \end{matrix}\right\} \qquad (B')$$
$$t^{\text{right}}$$
$$b^{\text{right}}$$

On this view new quarks will always come in pairs, one of charge $\frac{2}{3}$ (t) and one of charge $-\frac{1}{3}$(b). It is also expected that each column of (B) is matched by a lepton column like (A) of p. 112 so that (B′) would imply the existence of a new lepton τ (tau) with its associated neutrino ν_τ, giving:

$$\left.\begin{matrix} \tau^{\text{left}} \\ \nu_\tau^{\text{left}} \end{matrix}\right\} \qquad (A')$$
$$\tau^{\text{right}}.$$

A candidate for τ was at hand in an oxymoronic particle, the heavy lepton, of about twice the proton mass, which had been found at SPEAR. Its associated idiosyncratic neutrino ν_τ was later confirmed.

The theoretical reason for expecting the linked column structure of quarks and leptons exemplified by (A) and (B), and (A′) and (B′), lies in a need to cancel a type of divergence called an anomaly which would otherwise spoil the renormalizability of the theory. The cancellation of this disagreeable feature can be achieved by trading off the lepton anomaly against the quark anomaly, the factor of 3 from the colour degrees of freedom of the quarks playing a vital role in the balancing act. This would imply a deep interlocking between leptons and quarks, and has encouraged the search for super-unified theories which combine weak, electromagnetic, and strong interactions in a grand synthesis. A delicate point for such attempts is to come to terms with the fact that we have never seen baryons turning themselves into leptons (and so violating the conservation of baryon number). This means that quarks and leptons must retain a degree of distinctness from each other. So far theories of this kind are not much more than gleams in the eyes of the more speculatively minded theorists. It is also worth reminding ourselves that the anomaly linkage between quarks and leptons takes the criterion of renormalizability very seriously, a point of view about which we expressed reservations in Chapter V.

One thing is clear from looking at the arrays of (A), (A′), (B), and (B′). There is a mysterious prodigality in nature, the first hint of which we had when we asked in Chapter I, 'Who ordered the muon?'. The basic structure of the theory is sufficiently exemplified by the first columns of (A) and (B). Why is it duplicated again, twice over? No doubt there is an answer, but its discovery lies in the future. For the present, we seem to be faced with an extensive, almost plethoric, portfolio of elementary objects. There are the three sorts of leptons (e, μ, τ) and their associated neutrinos, making six particles in all. There are the three pairs of quarks (u, d; s, c; t, b) each existing in three colours, so that there are eighteen varieties altogether. Then there are the fundamental bosons, the photon and the three intermediate vector bosons for the mediation of the unified weak and electromagnetic interactions, and the gluons, eight in number, to bind the quarks together in hadronic matter. That seems to make 36 elementary particles in all, not to mention our elusive friend the graviton and some further esoteric particles called Higgs mesons to which we have only been able to give a glancing reference (p. 100). It is certainly not as economic as air, earth, fire, and water.

Marriage à la mode

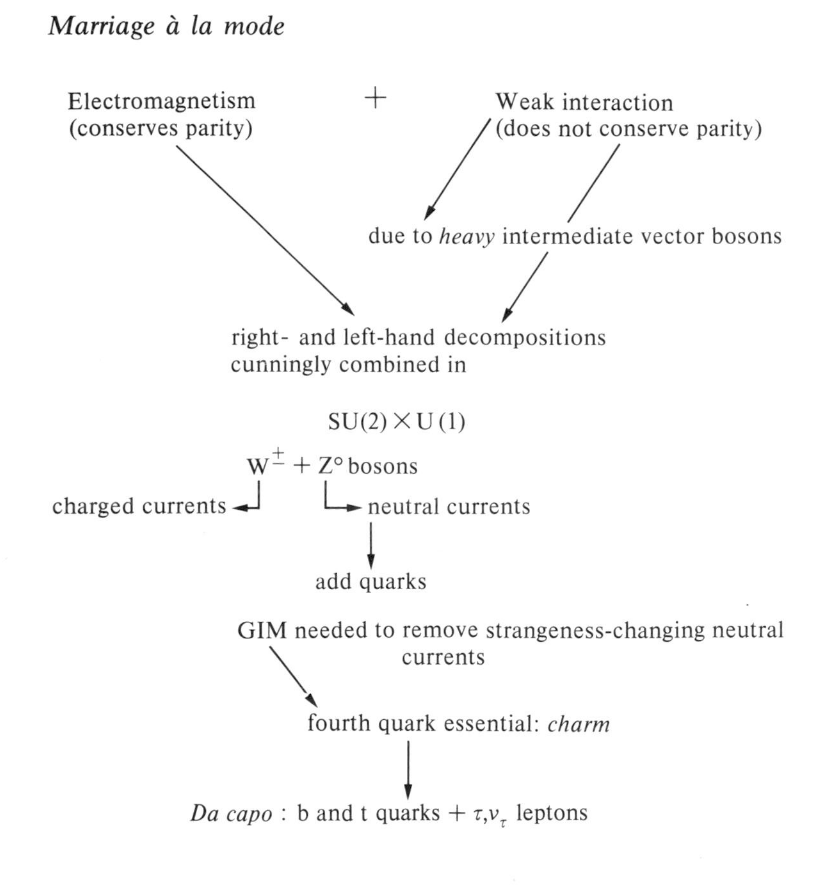

IX

Forthcoming Attractions

There is reason to hope that the heady progress that we have seen in particle physics in recent years will continue in the future. On the experimental side of the subject advances are likely to come from the opening of new facilities which will give access to regimes at yet higher energies.

In November 1978 an electron–positron colliding ring called PETRA came into operation at Hamburg. The energy it provides is four times that previously accessible in such a device. We have already noted (Chapters VII and VIII) that e^+e^- machines are particularly well suited to the study of many of the questions of greatest contemporary interest. It is hoped that PETRA may provide evidence for the expected t-quark, though, since no one can predict how heavy the t is, it must be a question of waiting and seeing as far as this is concerned. Certainly PETRA will provide sufficient energy for the weak interactions to become strong enough to start mixing with electromagnetic effects. The Salam–Weinberg theory will be put to further test. However, the crucial question of the existence of the Ws and the Z cannot be answered at PETRA. There is still not enough energy to create these very heavy particles.

The proton–antiproton collider at the CERN SPS (Chapter II) should be able to perform the task of producing intermediate vector bosons when it starts operating in the early 1980s. The experiment to find them will not be easy. At these very high energies many things happen and only a tiny fraction of the events would involve Ws or a Z. It will call for great skill and discrimination on the part of the experimentalists to winkle them out, always assuming that they are there. The discovery of intermediate vector bosons would consolidate much of our present understanding. But if it happened that they were not there after all, while the disappointment would be sharp, the stimulus

The Deutches Electronen Synchroton site at Hamburg. The ring for PETRA encircles the perimeter. (*photograph–DESY*)

to new thought would be considerable. In an experimentally-led subject there is a certain 'heads we win, tails they lose' aspect to what the future holds.

Although the proton–antiproton collider might well discover the Ws and Z, the detailed study of their properties would be best done with a Large Electron–Positron machine (acronymically LEP). There are plans afoot which, it is hoped, will result in the building of LEP at CERN. It would have an energy four or five times that of PETRA and be housed in an underground tunnel 10 km in diameter. It takes two or three years to plan such a machine and about seven years to construct it, so that LEP will not be on the scene before the end of the 1980s. Not only would it be the ultimate probe of the unification of electromagnetic and weak interactions but it would also enable us to continue the saga of quark flavour. Are there quarks yet heavier than the b and (expected) t? Do they indeed organize themselves into the column structure of Chapter VIII and are there leptons in attendance as we expect? If the reader has caught anything of the excitement of the subject he will see the importance of answering such questions. They are the questions that arise naturally today. If the past is any guide to the future, a most important aspect of coming experimentation is to be to open to the unexpected. However interesting may be the answers to the questions which we now know how to frame, how much more fascinating are likely to be the surprises in store.

The coming activity on the theoretical side will not be just a scramble to interpret the new experimental results as they become available. Important developments are taking place in our basic understanding of relativistic quantum mechanics. They centre round the discovery of certain classical solutions of gauge theories, called generically *instantons*, and the interpretation of their relevance for the quantum theory. This development offers a hope of getting away from a purely perturbation-theory-based understanding of quantum field theory. Once again it seems that relativistic quantum mechanics is exhibiting a richness of structure unexpected hitherto.

Have we really reached the centre of the elementary particle onion? Is confinement the signal of the ultimate constituents of matter? If so, why are there so many of them?

The play continues. The script, it seems, never ends with a full stop but only with a further question mark.

X

I am not an habitual reader of popular books on science but when I do peruse one I am from time to time irritated by the *obiter dicta* in which its author indulges. The subject is frequently religion, the tone inimical, but I do not think that my vexation is simply the pique of a Christian believer at hearing his faith disparaged. It seems to me that the comments are often shallow, sometimes ignorant, and their presence in the text which is otherwise an authorative exposition of a highly successful activity gives them an apparent weight which they do not in fact possess. The answer clearly is not counter-propaganda of the same kind. I have tried in writing this book to stick strictly to its theme and not give the reader the benefit of my views on wider issues. However, as a reward for my scrupulosity I am permitting myself the indulgence of these parting few words.

There are two things I would like to say. One is that, of course, there are many puzzles which relate to that North-West frontier of knowledge that can roughly be called the 'science and religion' borderline. To most of these we are not at present in a position to give answers. For example, it is an interesting question whether biology is physics writ large, in the sense in which chemistry is certainly physics writ large. In principle (though scarcely in practice), all chemical facts are deducible from physical first principles. Even more in principle, is the same true of biological systems or are they of a scale which introduces something new, not discernible in a catalogue of the separate parts? It seems to me that some biologists say one thing and some the other and that this means that we do not really know the answer. Even more fascinating, how is our experience of consciousness related to the physical events occurring in our brains? (It is curious to note that some people believe that consciousness is fundamental to the interpretation of quantum mechanics, which if it were so might cut quite a few Gordian knots. But the question is still very far from settled.)

These sorts of questions are of the highest interest but it is not helpful to suppose we know the answers to them before in fact we do so. In a period of enforced agnosticism about the answers to these questions, it seems to me important to hold fast to primary experience and not to subject it to a procrustean over-simplification in search of premature understanding. That is a waiting attitude which science itself encourages. The physicists of the early years of this century would not have made progress by denying or discarding the wave-like or the particle-like aspects of the photon. It was necessary to hold to both, in dialectical tension, for many years before quantum field theory dissolved the apparent paradox. So in our life as persons I believe we must hold fast to the insights of science, and to our experiences of beauty, choice (however circumscribed), and moral responsibility. And to that list I would want to add the worship of God and the knowledge of his grace in Jesus Christ.

The second thing I want to say is a brief word about is whether I detect any consonance between the world view of Christianity and the world view of science. In fact I do, but I do not suppose that I can prove it for you, any more than someone else could disprove it. We are in an area of discourse where such knock-down arguments are not available to anyone. I can only share an insight which comes to me.

I am very struck by the fact, first mentioned in Chapter I and continually illustrated by all that followed, that mathematics, which essentially is the abstract free creation of the human mind, repeatedly provides the indispensable clue to the understanding of the physical world. This happening is so common a process that most of the time we take it for granted. At root it creates the *possibility* of science, of our understanding the workings of the world. It seems to me a remarkable fact. I believe—I cannot prove it—that it is one aspect, perhaps rather a small one really, of the logos doctrine of Christianity. Israel developed an idea of the Word of God who was his agent in the creation of the world. The prologue to St. John's gospel not only makes the astonishing identification of that Word with Jesus of Nazareth but also says that the Word is the true light that lightens every man. The use of mathematics to comprehend the universe shows a relation between the workings of our minds and the structure of the world. I believe that this is one aspect of what the writer of the Fourth Gospel is telling us.

In the creation myth of Genesis, Adam is given lordship over nature in the naming of the animals. The pursuit of science is an aspect of the *imago dei.* Therefore it does not seem to me strange that these words which I have written while Professor of Mathematical Physics in the University of Cambridge will be published when I am an ordinand studying for the Anglican priesthood at Westcott House.

Glossary

This glossary does not seek to give precise definitions of technical terms, which would be as inaccessible for those who needed them as they would be otiose for those who could understand them. Rather, it tries to convey the character of the concept lying behind a word in a way which will help the layman to follow the text. Even such partial attempts at definition must be interlocking and terms treated in the glossary are italicized when used elsewhere.

analyticity: a powerful mathematical property expressing smoothness of behaviour, not just for physical values of the quantities involved but also when they are given imaginary parts proportional to the square root of -1. One can say that functions with this property are mathematically 'nice'. They have the 'global' property that knowledge of their values in one region determines their value elsewhere.

angular momentum: a quantity which assesses the amount of rotatory motion present. The larger it is, the harder it is to stop rotation.

antimatter: particles are divided into classes, matter and antimatter, which have opposite characteristics and are capable of mutually annihilating each other to produce energy. A particle's partner (in this sense) is called its *antiparticle.*

asymptotic freedom: the property that certain interactions become weak at short distances so that particles become almost free of the forces constraining them.

baryon: a *strongly interacting fermion.* The most common examples are *protons* and *neutrons.*

baryon number: the number of *baryons,* reckoned by assigning $+1$ to each baryon and -1 to each antibaryon (the *antiparticle* of a baryon). It is believed to be a conserved quantity. *Mesons* are conventionally assigned baryon number zero.

β-decay: the decay of a radioactive nucleus, or a particle, with the emission of an *electron.* Historically it is the prototype *weak interaction.*

bootstrap: the notion that no particles are elementary but they are all made out of each other.

Bose statistics: the *statistics* of particles which like to be in the same state.

boson: a particle obeying *Bose statistics.* By the *spin-and-statistics theorem* it must have integral *spin.* Well-known bosons include the *photon* and all kinds of *meson.*

bottomness: a *quantum number* associated with the b-quark and particles containing it.

cascade particles (Ξ)*: baryons* of *strangeness* -2 which undergo a chain of decays.

classical mechanics: a dynamical theory appropriate to circumstances in which the effects of *quantum mechanics* are negligible.

charge conjugation (C)*:* the operation of replacing particles by *antiparticles* and vice versa. For example a system of electrons and antiprotons is changed into a system of positrons and protons.

charged current: a *weak interaction* in which charge is exchanged between the participating particles (typically involving an *electron* or a *muon* changing into a *neutrino*, or vice versa).

charm: a *quantum number* associated with the c-quark and particles containing it.

colour: a quantum number which serves to differentiate three varieties of each type of *quark*. It also differentiates eight types of *gluon*. Colour is not manifested by observed particles, which are all 'white'.

configuration space: a description of particles in terms of where they are located.

confinement: the property that *quarks* and *gluons* cannot escape from the particles of which they are the constituents.

conservation law: the requirement that the total amount of some quantity (e.g. energy or electric charge) does not change. Due account must be taken of the sign, for example an electric charge of +1 can change into two +1s and one −1 without violating charge conservation. Conservation laws can be additive (different contributions add together; as for energy, momentum, etc.) or multiplicative (different contributions (±1) multiply together; as for *parity*, etc.).

cosmic rays: high energy particles incident on the earth from outside it.

coupling constant: a measure of the strength of a particular interaction which couples particles together. The best-known example is electric charge which determines the interaction of a particle with the photons of the electromagnetic field.

CP: the combined operation of *charge conjugation* and *parity reflection*, which takes one from the motion of a particle to the apparent motion of its antiparticle observed in a mirror.

cross-section: a measure of the likelihood that two particles will interact. It can be visualized as the size they present to each other, so that the larger the cross-section the more likely is a collision.

crossing: the interchange of particles between the initial and final states of a process, thereby giving a new process; more particularly, the property that a single function describes all the processes which can be related in this way.

cyclotron: one of the earliest forms of particle accelerator.

doublet: a pair of particles with similar properties.

electromagnetic force: one of the basic forces of nature, associated with electrically charged particles and due to the exchange of *photons*. Electromagnetic forces hold atoms together and cause them to interact with each other, thereby giving the large-scale properties of matter. *Electromagnetic interactions* result from experiencing this force.

electron: a light negatively charged particle. It was the first elementary particle to be discovered.

electron number: the number of *electrons* and electron-type *neutrinos.* It is believed to be a conserved quantity.

exclusion principle: the property that there can only be at most one *fermion* in each possible state of motion.

Fermi statistics: the *statistics* of particles whose behaviour is such that there is never more than one in each state.

fermion: a particle obeying *Fermi statistics.* By the *spin-and-statistics theorem* it must have half-odd integral spin. Prominent examples of fermions are *electrons, protons,* and *neutrons.*

Feynman integrals: neat mathematical expressions for the terms in *perturbation theory.*

flavour: a generic name for *quantum numbers* like *isotopic spin, strangeness,* etc. which are manifested both by *quarks* and also by observed particles. It contrasts with *colour,* which is only manifested by quarks and gluons.

Fourier analysis: a mathematical technique for analysing vibrating systems into their component frequencies. The analysis of a musical note in terms of the fundamental and its harmonics is an example.

four-momentum: momentum and energy combined to make a *four-vector.*

four-vector: a *vector* in four-dimensional space-time which provides a natural mathematical setting for *relativity.* Examples are provided by the combination of spatial position and time, and by *four-momentum.*

gauge boson: a *spin*-1 particle associated with a *gauge theory.*

gauge theory: a highly symmetrical form of *quantum field theory.* Such theories are tightly knit in their construction and have many attractive properties. They provide the currently most favoured formalism with which to discuss basic particle theory.

(Historical note: The name is far from self-explanatory and in fact an interesting story lies behind it.

Hermann Weyl wished to find a way of making electromagnetic interactions expressible in terms of geometry, as Einstein had succeeded in doing for gravity in his general theory of relativity. It turned out that Weyl could make the generalisation provided he assumed a theory in which vectors changed their lengths as they were moved around. This was interpreted as meaning that there was no intrinsic length scale so that arbitrarily chosen scales could be changed locally at will. These local changes of scale were called gauge (= scale) transformations and they induced on the electromagnetic part of the geometry just those transformations which express that symmetry of electromagnetism which we now call gauge invariance.

Weyl's idea collapsed almost as soon as he invented it because it was incompatible with quantum theory. Planck's constant combined with the velocity of light and the mass of the electron yields an absolute length scale given by the Compton wavelength of the electron, $\hbar/mc$. The baby was gone but the bath water remained in the form of a retention of the terminology of gauge transformations).

GIM mechanism: a cancellation mechanism which removes the unwanted (because experimentally not present) prediction of *strangeness*-changing *neutral current* events.

gluon: the *flavour*less particles whose exchange creates the forces between *quarks*. They are thought to be the *gauge bosons* of the *colour gauge theory*.

Goldstone boson: an unwanted zero mass particle which would be associated with *gauge theories* with *spontaneous symmetry breaking* were it not removed by a device invented by Higgs.

graviton: the hypothetical particle whose exchange mediates the gravitational force.

gyromagnetic ratio: the ratio of a particle's *magnetic moment* to its *spin*.

hadron: a particle subject to the *strong force*.

hadron spectroscopy: the classification of *hadrons* into sets (*multiplets*) of particles with similar properties, together with a specification of those characteristics by which members of a multiplet differ from each other.

handedness: the circumstances that some property of a system can only be stated by using a right- (or left-)hand rule in order to specify a direction. Such a rule is altered by reflection (which interchanges right and left).

hypercharge: a *quantum number* compounded of *strangeness* and *baryon number*.

instantons: finite energy solutions of classical *gauge theories*. Their significance for the interpretation of quantum gauge theories is at present the subject of intense investigation.

intermediate vector boson: a generic name for the particles ($W^{\pm}$, Z^0) thought to mediate the *weak force*.

intersecting storage ring: a device in which rapidly moving beams of particles can be stored and made to collide with each other. In particular, the *ISR*, a device of this kind at CERN in which two beams of protons collide. It currently affords the highest energy regime accessible in the laboratory.

intrinsic quantum number: see *quantum number*

invariance: the property of remaining unaltered when other quantities are changed. Such behaviour results from the system possessing a symmetry. For example, a circle is invariant under rotations about its centre. If the change can be made gradually in small steps (as for rotations) the symmetry is called continuous; if the change must be made at one go (as for reflections) the symmetry is called discrete.

isotopic spin: an important symmetry possessed by *hadrons*. Its existence was first noted when it was observed that *protons* and *neutrons* behave similarly under the *strong force*. More generally isotopic spin is responsible for the existence of *multiplets* of hadrons which differ from each other only by their *electromagnetic* and *weak interactions*.

kaon (*K-meson*): a type of *strange meson*.

lepton: a *fermion* with no *strong interaction.*

Lie group: a mathematical construction which is associated with *invariance* under a continuous symmetry. A simple physical realization of a Lie group is provided by the set of rotations in three-dimensional space.

linear accelerator: a long straight device for accelerating particles.

magnetic moment: the measure of how a particle behaves in its *electromagnetic interactions* as if it were a small magnet.

mass shell: the condition, which must be satisfied for physical particles, that the squares of their *four-momenta* equal the squares of their masses.

matrix: a mathematical quantity in the form of a rectangular array of numbers. Matrices can be multiplied together and used to give *representations* of *Lie groups.*

meson: a *hadron* which is a *boson.* Examples of mesons are *pions* and *kaons.*

momentum space: a description of particles in terms of their momenta.

multiplet: a set of particles with similar properties.

muon (μ-meson): a *lepton* which is to all intents and purposes a heavier version of the *electron.*

muon number: the number of *muons* and muon-type *neutrinos.* It is believed to be a conserved quantity.

neutral current: a *weak interaction* in which there is no exchange of charge between the participating particles (typically, neutrino scattering). It is to be contrasted with a *charged current* interaction.

neutrino: a massless neutral *lepton* which has only *weak interactions.* There are several types of neutrino associated respectively with *electrons, muons,* and *tau particles.*

neutron: a neutral *baryon* which is one of the constituents of nuclei.

nucleon: a generic name for *protons* and *neutrons,* the constituents of nuclei.

parity (P): the operation of going from the direct observation of a system to the study of its reflection in a mirror. More specifically, parity is a quantity (taking values ± 1) associated with the behaviour of systems under reflection.

particle–wave duality: the fact that microscopic systems behave as if they were particles and also as if they were waves. This was at first considered paradoxical, but is now completely understood in terms of *quantum field theory.*

parton model: a picture of the observed *hadrons* as being composed of more elementary constituents (*partons*).

Pauli exclusion principle: see *exclusion principle.*

periodic table: the arrangement of the chemical elements which exhibits certain regular recurrences in their behaviour.

perturbation theory: an approximate method of calculation. The effect being studied is divided into two parts, a large one which is easy to treat exactly and a small one where the difficulty of calculation resides (the perturbation). The answer is then obtained as a mathematical series in powers of a small parameter specifying the size of the perturbation.

phase space: the set of final state configurations which are possible in an interaction. The larger this set, the more frequently that interaction will occur.

phonon: the particle aspect of sound waves in solids.

photon: the *vector particle* which mediates the *electromagnetic force.*

pion (π*-meson*): the lightest *meson.* Its exchange is responsible for almost all the *strong forces* in nuclei. Pions are found with electric charges $+1, 0, -1$.

positron: the positively charged *antiparticle* of the *electron.*

proton: a positively charged *baryon* which is one of the constituents of nuclei.

quantum chromodynamics (*QCD*): the putative fundamental theory of *strong interactions* based on a *gauge theory* of the *colour* properties of *quarks* and *gluons.*

quantum electrodynamics (*QED*): the theory of the *electromagnetic interactions* of *electrons*, *muons*, and *photons.*

quantum field theory: the formalism obtained by applying *quantum mechanics* to a field, that is a quantity which varies from point to point. Quantum field theory provides the fundamental formulation of *relativistic quantum mechanics.*

quantum mechanics: the principles of mechanics essential for the description of microscopic systems for which Heisenberg's *uncertainty principle* must be taken into account.

quantum number: an intrinsic property possessed by every particle of a specific type. Typically it is measured in integer units. Examples are electric charge, *baryon number, strangeness,* etc.

quark: a *fermion* which is believed to be one of the fundamental constituents of matter. Different types of quark correspond to different *flavours* and *colours.* All quarks have fractional electric charge.

relativistic effects: Consequences, such as increasing mass and slow running internal clocks, which manifest themselves as particles acquire velocities which are an appreciable fraction of the velocity of light.

relativistic quantum mechanics: the theory, essential for the description of small fast-moving particles, which combines *quantum mechanics* and *relativity.*

relativity (strictly, the special theory of relativity): The principles of mechanics essential for the description of particles whose velocities are an appreciable fraction of the velocity of light.

renormalization: the necessary addition to the mass, charge, etc. of a particle of the effects due to its interaction. More particularly, the procedure whereby certain *quantum field theories* can be made to yield finite sensible results by identifying the total mass, charge, etc. with experimentally determined values.

representation: a mathematical picture of a *Lie group.*

resonance: a very unstable particle; the enhancement in the *cross-section* due to the presence of that unstable particle as an intermediate state.

scattering: the interaction of two particles which collide. As a consequence their motions can be deviated and additional particles may be produced.

singlet: a *multiplet* consisting of one particle only (that is, there are no other particles with similar properties).

S-matrix theory: the attempt to discuss *relativistic quantum mechanics* solely in terms of the relation between the initial and final states of *scattering* experiments.

spin: the intrinsic *angular momentum* possessed by a particle. It is measured in integral $(0, 1, 2, \ldots)$ or half-odd-integral $(\frac{1}{2}, \frac{3}{2}, \ldots)$ multiplets of Planck's constant, $\hbar$.

spin-and-statistics theorem: the requirement that particles of integral (half-odd-integral) *spin* are necessarily *bosons* (*fermions*).

spontaneous symmetry breaking: the process by which the solutions of a theory are less symmetrical than the theory itself.

statistics: the way particles behave in aggregate.

strangeness: an intrinsic *quantum number* associated with the group *SU(3)* and specifically with the s quark and particles containing it.

strong force: one of the basic forces of nature. As its name suggests, it is the strongest. The strong force holds *protons* and *neutrons* together in nuclei. *Strong interactions* result from experiencing this force.

SU(2): a *Lie group* whose representations can give the *multiplets* of *isotopic spin.* Consequently it is frequently used as a synonym for isotopic spin. SU(2) in that sense can be thought of as involving the shuffling of the u and d quarks only.

SU(3): a larger *Lie group* whose *representations* give larger *multiplets* of particles. SU(3) can combine within itself *isotopic spin* and *strangeness.* It can then be thought of as shuffling the u, d, and s quarks.

SU(N): a *Lie group* associated with shuffling *N* types of quark.

symmetry principle: the property that a theory does not change when certain operations are performed on it. Examples are symmetry under change of position, under rotations, under reflection, under interchange of particle and antiparticle, etc. (see also *invariance*).

synchrotron: a modern machine for accelerating particles.

synchrotron radiation: radiation, with consequent loss of energy, by charged particles whose motion is deflected by a magnetic field.

tau particle (τ)*:* a recently discovered charged heavy *lepton.*

TCP: the combined effect of *time reversal, charge conjugation,* and *parity.* This operation takes one from the direct observation of a particle to the observation of a film of the antiparticle, run backwards, and viewed in a mirror.

TCP theorem: the requirement of *relativistic quantum mechanics* that the two behaviours connected by *TCP* are identical.

time reversal (*T*)*:* the operation of reversing all motions. It is equivalent to running a film backwards, hence the name.

topness: a *quantum number* associated with the t-quark and particles containing it.

triplet: three particles with similar properties.

uncertainty principle: one cannot both know where a particle is and what it is doing. More specifically there is a limit to the accuracy with which position and momentum can both be measured, because measuring one alters the other in a not completely controllable way.

unitarity: the principle that the set of probabilities associated with the different final states which can result from *scattering* from a given initial state must all be positive and add up to one (that is, something must happen!)

vacuum fluctuations: the surprising fact that there are fluctuating force fields even when no particles are present. They result from the *zero-point motion* of *quantum field theory.*

vector: a quantity with both magnitude and direction. If the direction indicates a direction 'along', the vector is *polar,* if it indicates a direction 'around', it is *axial.*

vector particle: a *boson* of *spin* 1.

virtual particle: an exchanged object mediating a force but not itself an observable particle. (That is, it is not on the *mass shell.*)

weak force: one of the fundamental forces of nature. It is associated with the β-decay of nuclei and the decays of many particles. *Weak interactions* result from experiencing this force.

Weinberg angle: a parameter in the simplest theory of unified *electromagnetic* and *weak interactions.*

Yang–Mills theory: a *gauge theory* of *vector particles.*

zero-point motion: the 'quivering' which results from the fact that the *uncertainty principle* will not permit an oscillating system ever to be at rest.

Zweig's rule: an empirical rule that particles composed of a quark and antiquark of the same *flavour* (*strangeness, charm,* etc.) predominantly decay into particles which have that flavour.

Index

accelerators 25f
additive conservation law 42
analyticity 86f, 127
Anaximenes 1
angular momentum 6, 38, 65, 127
annihilation 20, 31, 103, 115
anomalous gyromagnetic ratio 77
anomalous magnetic moment 77
antimatter 16f, 19f, 127
antiparticle 19f, 85, 127
antiproton 19
arrow of time 45
associated production 51
asymptotic freedom 101, 105, 127
atom 2, 95
axial vector 41

bag model 96, 102
barn 53
baryon 13, 24, 65, 127
baryon number 51, 59, 118, 127
BEBC 32
β-decay 14, 127
Bohr, N. 2
bootstrap 89f, 127
Bose statistics 7, 127
bosons 7, 24, 127
bottomness 117, 127
b-quark 117, 120
bubble chamber 32f, 52

cascade particle (Ξ) 58, 60, 128
causality 87f
Čerenkov counter 33
CERN (Conseil Européen pour la Recherche Nucléaire) 26f, 108, 121, 123
Chadwick, J. 9
charge conjugation 44, 46, 128
charge conjugation parity 44
charge conservation 38
charge independence 11, 22, 57, 100
charge renormalisation 76
charged current 34, 110, 120, 128
charge-exchange reaction 29
charm 114, 116f, 120, 128
Chew, G. 89
Christianity 125
classical mechanics 128
clocks, moving 79
cloud chamber 19, 31
cobalt-60 39f
colour 103, 118, 128
compound nucleus 52
Compton, A. H. 5
Compton wavelength 129
configuration space 82, 128
confinement 95f, 102, 104, 128
conservation laws 36f, 49f, 128
continuous symmetry 42
cosmic rays 19, 49, 65, 128
coupling constant 51, 100, 128
CP 46f, 128
c-quark 113, 116
Cronin, J. W. 47
crossing 84f, 87, 128
cross-section 53, 128
cyclotron 27, 128

de Broglie, L. 5
decuplet 61, 65, 66
deep inelastic scattering 92f, 95, 105
delta-particle (Δ) 52, 61, 102f
detectors 25, 31f
Dirac, P. A. M. 6, 16f
discrete symmetry 42
DORIS 117
d-quark 64, 66, 104, 113

Eddington, A. S. 75
eightfold way 60
Einstein, A. 4, 12, 69, 80
electromagnetic force (electromagnetic interaction) 10f, 12, 15, 16, 22, 24, 25f, 39, 44, 45, 50, 51, 75, 88, 92, 98, 106, 120, 128
electron 3, 5, 7, 8, 11, 17f, 23, 24, 76, 92, 104, 111, 129
electron number 111, 129
electronic detectors 33, 35
entropy 45
eta-particle (η) 60
exclusion principle 7, 18, 129

Faraday, M. 17, 106
fermi 10
Fermi statistics 7, 129
Fermilab (Fermi National Accelerator Laboratory) 27, 117
fermions 8, 24, 129
Feynman, R. P. 23, 76, 78
Feynman integrals 78f, 86, 88, 107, 129

field theory 73
fine structure constant (α) 75
Finnegan's Wake 64
fission 10
Fitch, V. 47
flavour 104, 129
force 13f, 107
form factor 88
four-dimensions 80
Fourier analysis 70, 73, 74, 129
four-momentum 80, 82, 129
four-vector 80, 129
function 86
fundamental representation 63

Gargamelle 34
gauge boson 98f, 101, 129
gauge invariance 38, 129
gauge theory 51, 78, 98f, 101, 104, 109, 112, 123, 129
Gell–Mann, M. 49f, 60, 62, 64,
GIM mechanism 113, 120, 130
Glaser, D. 33
Glashow, S. 113
gluon 95, 102f, 104, 119, 130
Goldstone boson 100, 130
graviton 15, 130
gravity 14f, 16, 77
group 67
gyromagnetic ratio 17, 130

hadron 12, 16, 24, 49, 102, 130
hadron spectroscopy 66, 95, 104, 130
handedness 39, 108, 112, 130
Heisenberg, W. 13, 57, 83
Higgs meson 119
Higgs, P. W. 100
'hole' theory 19
hybrid device 33
hypercharge 59, 61, 64, 130

Iliopoulos, J. 113
inclusive experiment 93
infinities 76
infrared slavery 102
instanton 123, 130
interference phenomena 5
intermediate vector boson 106, 109f, 112, 120, 121, 130
intersecting storage ring 29, 130
interval 80
invariance 36f, 130
isobaric spin 57
isomorphism 56
isotopic spin 54f, 56, 64, 66, 96, 100, 130
ISR 29f

Joyce, J. 64
J/ψ particle 116f
J-resonance 115

kaon (K-meson) 44, 49f, 60, 113, 116, 130
Kemmer, N. 22

Lamb shift 77
lamda-particle (Λ) 51, 58, 60
Landau, L. D. 115
Lederman, L. 117
Lee, T. D. 39, 43f, 46
LEP 123
lepton 14, 16, 24, 107, 131
Lie group 55f, 62, 66f, 101, 131
lifetime 9, 21, 22, 23, 43, 49, 51, 52, 79, 116
light, nature of 5
linear accelerator 27, 131
logos doctrine 125

Maiani, L. 113
magnetic moment 17, 131
mass 9, 12, 19, 20, 23, 52f, 61, 66, 76, 81
mass shell 83, 87, 107, 131
matrix 56, 83, 131
Maxwell, J. C. 5, 106
Mendeleef, D. I. 2
meson 13, 14, 20, 24, 65, 131
Mills, R. L. 98
Millikan, R. A. 65
mirror nuclei 11
momentum space 82, 131
multiplet 13, 54f, 66, 131
multiplicative conservation law 42
muon (μ-meson) 21, 22, 24, 77, 79, 104, 106, 110, 111, 131
muon number 111, 131

neutral current 110f, 113, 120, 131
neutrino 9, 11, 14, 20, 24, 33, 35, 94, 109, 110, 111, 112, 118, 131
neutron 9, 10, 12, 22, 131
negative energy solutions 18
Newton, I. 5, 13, 80

nuclear democracy 89
nuclear force 10, 20
nucleon 13, 57f, 60, 131
nucleus 2, 94, 95

octet 60, 65, 66
Omega-minus (Ω^-) 60f
Oppenheimer, J R. 19
oscillator 70f, 74

pair creation 18, 20
parity 43, 44, 46, 112, 131
particle interpretation 73
particle-wave duality 5, 73, 125, 131
parton model 94, 131
Pauli, W. 8, 39
periodic table 2, 13, 131
perturbation theory 75, 78, 84, 86, 105, 123, 132
PETRA 121
phase space 45f, 132
phi-particle (ϕ) 116
phonon 79, 132
photoelectric effect 4
photon 4, 7, 8, 9, 11, 13, 16, 18, 19, 20, 24, 33, 45, 72, 73, 74, 92f, 98, 116, 125, 132
pion (π-meson) 21, 45, 60, 89, 103, 111, 113, 132
Planck, M. 4
Planck's constant ($\hbar$) 6, 71, 129
polar vector 41
positron 19, 132
probability 83, 114
proportional wire chamber 33
proton 4, 7, 8, 10, 12, 22, 92f, 132
PS 31
ψ-resonance 116
ψ'-particle 117

quantum chromodynamics (QCD) 89, 105, 132
quantum electrodynamics (QED) 75, 92, 101, 132
quantum field theory 6, 18, 73, 74, 84, 87, 125, 132
quantum number 4, 43, 50, 132
quantum theory 4, 18, 69, 70f, 74, 83, 125, 132
quark 63f, 66, 92f, 94, 95f, 103f, 113f, 120, 132
quark model 65f, 104

reflections 38f
relativistic effects 8, 9, 69, 79, 132,
relativistic quantum mechanics 8, 69f, 87, 96, 123, 132
relativity, general 78
renormalisation 75f, 77, 101, 118, 133
representation (of a group) 55, 67f, 133
resonance 52f, 72, 89, 116, 133
rho-meson (ρ) 89f
Richter, B. 115
Rosenfeld, L. 57
rotations 38, 55f, 66f
Rutherford, E. 2, 94

Salam, A. 108, 110
scattering experiments 25, 52f, 82f, 133
Schwinger, J. 76
scintillation counter 33
'sea' of negative energy electrons 18
second law of thermodynamics 45
Shaw, R. 99
sigma-particle (Σ) 51, 58, 60
simultaneity 80
singlet 65, 133
S-matrix 83f, 87f, 105, 133
space-like 80
space-time 80, 82
spark chamber 33
SPEAR 115f, 118
spin 6, 7, 17, 41, 57, 61, 65, 98, 102, 133
spin-and-statistics theorem 8, 79, 103, 133
spontaneous symmetry breaking 99f, 104, 133
SPS 26f, 31, 108, 121
s-quark 64, 66, 104, 113, 116
Stanford (Linear Accelerator Center) 27, 93, 115
statistics 7, 133
s, t variables 85
strangeness 50f, 57, 58f, 66, 113, 133
string model 96, 102
strong force (strong interaction) 10f, 15, 21, 24, 39, 44, 45, 50, 52, 58, 84, 133
SU(2) 56, 58, 66, 133
SU(3) 57f, 59f, 63f, 66, 101, 104, 133

supermultiplet 58
super-unified theory 118
symmetry principles 37f, 48, 54f, 98, 133
synchrotron 27, 133
synchrotron radiation 27, 133

τ-meson 43, 49
τ-particle 118, 120, 134
τ-θ puzzle 43f
TCP 47, 134
TCP theorem 47, 79, 134
θ-meson 43, 49
Thomson, J. J. 3
time-like 80
time reversal 44f, 46, 134
Ting, S. 115
Tomonaga, S. 76
topness 117, 134
t-quark 117, 120, 121
translation 36f

uncertainty principle (uncertainty relation) 52f, 71f, 82, 93, 134
unified theory of weak and electromagnetic interactions 106f, 108, 112, 121
unitarity 84, 87, 134
u-quark 64, 66, 102, 104, 113
upsilon-particle (Υ) 117

vacuum 73
vacuum fluctuations 74, 75f, 134
vector 40, 134
virtual particle 107, 134

wave-particle duality 5, 73, 125, 131
W-boson 107, 109, 120, 121
weak force (weak interaction) 14, 16, 21, 23, 24, 34, 39, 43, 44, 46, 47, 50, 51, 58, 62, 94, 106, 134
Weinberg angle 109, 111, 134
Weinberg, S. 108, 110
Westcott House 126
Weyl, H. 129
Wilson, C. T. R. 31

Xi-particle (Ξ) see cascade particle
Ξ^*-particle 61

Yang, C. N. 39, 43f, 46, 98
Yang–Mills theory 98f, 134
Young, T. 5
Y*-particle 61
Yukawa, H. 20f

Z-boson 110f, 120, 121
zero-point motion 72f, 134
Zweig's rule 117, 134